FATIGUE DESIGN
OF COMPONENTS

T0349045

Other titles in the ESIS Series

For information on how to order titles 1–21, please contact MEP Ltd, Northgate Avenue, Bury St Edmonds, Suffolk, IP32 6BW, UK

FATIGUE DESIGN
OF COMPONENTS

Editors: G. Marquis and J. Solin

ESIS Publication 22

This volume represents a selection of papers presented at the Second International Symposium on Fatigue Design, FD' 95, held in Helsinki, Finland on 5–8 September, 1995. The meeting was organised by VTT Manufacturing Technology and co-sponsored by the European Structural Integrity Society (ESIS) and the American Society for Testing and Materials (ASTM). Partial funding for the event was provided by the Commission of the European Communities and City of Helsinki.

Elsevier

Amsterdam • Oxford • New York • Lausanne • Tokyo • Shannon

AMSTERDAM	Elsevier Science BV, Sara Burgerhartstraat 25, 1055 KV Amsterdam, The Netherlands
U.K.	Elsevier Science Ltd, The Boulevard, Langford Lane, Kidlington, Oxford, OX5 1GB, U.K.
U.S.A.	Elsevier Science Inc., 660 White Plains Road, Tarrytown, New York 10591-5153, U.S.A.
LAUSANNE	Elsevier Science SA, P.O. Box 564, CH-1001 Lausanne, Switzerland
JAPAN	Elsevier Science Japan, Higashi Azabu 1-chome Building 4F, 1-9-15 Higashi Azabu, Minato-ku, Tokyo 106, Japan
SHANNON	Elsevier Science Ireland Ltd, Bay I5K, Shannon Industrial Estate, Co Clare, Ireland

First Edition 1997

Library of Congress Cataloguing in Publication Data
A catalog record for this book is available from the
Library of Congress

British Library Cataloguing in Publication Data
A catalogue record for this book is available from the
British Library

ISBN 0 08 043318-9

In order to make this volume available as economically and as rapidly as possible the authors' typescripts have been reproduced in their original forms. This method unfortunately has its typographical limitations but it is hoped that they in no way distract the reader.

Transferred to digital printing 2005
Printed and bound by Antony Rowe Ltd, Eastbourne

SYMPOSIUM ORGANISERS

Scientific Committee

S. Berge	*Norway*	Y. Murakami	*Japan*
A. Blom	*Sweden*	E. Niemi	*Finland*
G. Glinka	*Canada*	J. Petit	*France* / ESIS
P. Heuler	*Germany*	J. Polák	*Czech Republic*
M. Mitchell	*USA* / ASTM	D. Socie	*USA*
K. Miller	*UK*	R. Sunder	*India*

Advisory Board

H. Agerskov	Technical Univ. of Denmark, *Denmark*
D. Allen	European Gas Turbines, *UK*
A. Bakker	Delft Univ. of Technology, *The Netherlands*
C. Berger	Siemens AG, *Germany*
A. Bignonnet	Peugeot SA, *France*
G. Cavallini	University of Pisa, *Italy*
T. Dahle	ABB Corporate Research, *Sweden*
J. Devlukia	Rover Group, *UK*
A. Diboine	Renault, *France*
K-O. Edel	Fachhochschule Brandenburg, *Germany*
A. Gusenkov	Mechanical Engineering Research Inst., *Russia*
P. Haagensen	The Norwegian Inst. of Technology, *Norway*
G. Härkegård	ABB Power Generation, *Switzerland*
H. Jakubczak	Warsaw Univ. of Technology, *Poland*
B. Johannesson	Volvo AB, *Sweden*
A. Kuisma	Finnish State Railways, *Finland*
J. Llorca	Technical Univ. of Madrid, *Spain*
F. Luiz Bastian	COPPE/UFRJ, *Brazil*
T. Mizoguchi	Kobe Steel, *Japan*
C. Moura Branco	Instituto Superior Téchnico, *Portugal*
D. Morrow	Manta Corporation, *USA*
V. Panasyuk	Academy of Sciences, *Ukraine*
R. Rabb	Wartsila Diesel International, *Finland*
K. Rahka	VTT Manufacturing Technology, *Finland*
I. Riikonen	Kone Cranes, *Finland*
J. Samuelsson	VME Industries, *Sweden*
R. Stickler	Univ. of Vienna, *Austria*
T. Yoshimura	Toyota Motor Corporation, *Japan*

Local Organisers (VTT Manufacturing Technology)

Gary Marquis, Chairman
Jussi Solin, Co-chairman
Hilkka Hänninen
Sari Karvonen
Åsa Åvall
Kari Hyry (TSG-Congress Ltd.)

Elsevier Titles of Related Interest

Books

ABE & TSUTA
AEPA '96: Proceedings of the 3rd Asia-Pacific Symposium on Advances in Engineering Plasticity and its Applications (Hiroshima, August 1996). ISBN 008 042824-X

BAI & DODD
Adiabatic Shear Localization. ISBN 008-041266-1

BILY
Cyclic Deformation and Fatigue of Metals. ISBN 0-444-98790-8

CARPINTERI
Handbook of Fatigue Crack Propagation in Metallic Structures. ISBN 0-444-81645-3

KARIHALOO ET AL.
Advances in Fracture Research: Proceedings of the 9th International Conference on Fracture (Sydney, April 1997). ISBN 008 042820-7

KISHIMOTO ET AL.
Cyclic Fatigue in Ceramics. ISBN 0-444-82154-6

KLESNIL & LUKÁŠ
Fatigue of Metallic Materials. 2nd Edn. ISBN 0-444-98723-1

LÜTJERING & NOWACK
Fatigue '96: Proceedings of the 6th International Fatigue Congress (Berlin, May 1996). ISBN 008-042268-3

MENCIK
Strength and Fracture of Glass and Ceramics. ISBN 0-444-98685-5

PANASYUK ET AL.
Advances in Fracture Resistance and Structural Integrity (ICF 8). ISBN 008-042256-X

Journals

Acta Metallurgica et Materialia
Composite Structures
Computers and Structures
Corrosion Science
Engineering Failure Analysis
Engineering Fracture Mechanics
International Journal of Fatigue
International Journal of Impact Engineering
International Journal of Mechanical Sciences
International Journal of Non-Linear Mechanics
International Journal of Solids and Structures
Journal of Applied Mathematics and Mechanics
Journal of the Mechanics and Physics of Solids
Materials Research Bulletin
Mechanics Research Communications
NDT & E International
Scripta Metallurgica et Materialia
Theoretical and Applied Fracture Mechanics
Tribology International
Wear
Welding in the World

For more information Elsevier's catalogue can be accessed via the internet on http://www.elsevier.nl

Contents

PREFACE

This volume represents a selection of papers presented at the Fatigue Design 1995 Symposium held in Helsinki, Finland. The meeting was organised by VTT Manufacturing Technology and co-sponsored by the European Structural Integrity Society (ESIS) and the American Society for Testing and Materials (ASTM). FD'95 was planned following positive response to the previous Fatigue Design 1992 symposium and resulting special ESIS publication (*Fatigue Design*, ESIS 16).

One objective of the Fatigue Design symposium series is to help bridge the gap that sometimes exists between researchers and engineers responsible for designing components against fatigue failure. The large portion of papers authored by engineers working in industry illustrates that this objective is being realised. Papers provide an up-to-date survey of engineering practice and a preview of some design methods that are advancing toward routine application.

Subject areas touch many fields of engineering: automotive and ground vehicle industries, structural engineering, and power generation. It is our belief that such multi-disciplinary forums help stimulate the transfer of technologies from one field to the next and between academia, researchers and practising engineers. A majority of the papers deal with testing, design and analysis of real components. The problems are real and the methods developed and applied can hopefully provide new ideas for others responsible for ensuring adequate fatigue strength for components and structures.

Readers will find several key themes which arise from these proceedings. First is the increasingly important role that multiaxial fatigue continues to play in component design, testing and analysis. A related theme is the continued move toward fatigue testing and design using more realistic load histories. Uniaxial constant amplitude load testing is being supplimented by spectrum and multiaxial load testing. The third significant theme is the developing role of reliability methods in fatigue design. Innovative designs cannot rely on previously determined safety factors. Optimisation demands that variability in material properties, component usage and manufacturing be considered. The final key theme is the importance of introducing design rules that are easy to apply. Design methods must always provide a balance between accuracy and simplicity appropriate for the industry and type of component being considered.

A large number of people contributed to the success of both the symposium and this publication. The editors gratefully acknowledge the roles played by the scientific committee, manuscript reviewers, advisory board, as well as the staff members at ESIS, Elsevier, ASTM, and VTT. The most significant effort and greatest thanks, however, is due the authors who have invested countless hours both in the laboratory and in preparing their papers.

G. Marquis and J. Solin, Editors

FATIGUE ASSESSMENT OF AN AUTOMOTIVE SUSPENSION COMPONENT USING DETERMINISTIC AND PROBABILISTIC APPROACHES

J. Devlukia, Rover Group, Gaydon Test Centre, Lighthorne, UK
H. Bargmann, I. Rüstenberg, EPF Lausanne, Lausanne, Switzerland

ABSTRACT

This paper contains a case study on the fatigue assessment of a forged component for an automotive suspension. The fatigue strength reduction effects of the 'as-forged' surface resulting from the surface roughness and the presence of residual stresses were investigated. Test results under constant and variable amplitude bending loads are presented. In addition, test data on hourglass specimens specially prepared from the component material (a low alloy, medium carbon steel) under two different surface conditions, were also obtained to compliment the fatigue studies for the suspension arm.

A range of analytical techniques were used to predict the effects of the surface condition on the fatigue behaviour, including a probabilistic approach in which the surface roughness and residual stresses were treated as random variables.

KEY WORDS

Suspension arm, surface condition, roughness, residual stresses, fatigue, fatigue assessment, forged component, probabilistic approach

NOMENCLATURE

b, c	Fatigue strength, ductility exponents
E	Modulus of elasticity
K_t	Theoretical stress concentration factor
K_f, K_{sf}	Fatigue notch factor, surface strength reduction factor
k', n'	Cyclic strength coefficient, strain hardening exponent
$N_f, 2N_f$	Fatigue life: number of cycles, number of reversals
R	Load (stress) ratio
r_a, r_t	Average, peak surface roughness
R_m	Ultimate tensile strength
ΔS	Nominal stress range
$\Delta\varepsilon, \Delta\sigma$	Local strain range, stress range
ε_a, σ_a	Strain amplitude $\Delta\varepsilon/2$, Stress amplitude $\Delta\sigma/2$

1

$\varepsilon'_f, \sigma'_f$ Fatigue ductility, strength coefficient

σ_m, σ_r Mean, residual stress

$F_{\underline{N}}(N)$ Fatigue-life distribution

INTRODUCTION

The automotive industry is under continuous commercial and social pressure to deliver lower cost, more reliable and more environmentally friendly products for mass consumption. Hence, the challenges faced by the design engineers in this industry include delivery of lightweight but durable structures, reliability in service and shorter design lead times. In turn, these challenges generate requirements for tools and methodologies which allow optimisation of load bearing components for minimum weight while guaranteeing failure free performance in service. In other words, the ability to realistically predict with accuracy the fatigue life of a production part under complex service loading conditions is vital for the delivery of efficient designs.

Fatigue design methodologies for metallic components in the ground transport industries have been traditionally based on the so called Stress-Life (S-N) curve credited to Wöhler. This approach essentially deals with linear elastic stresses and strains, and hence it is applicable to components such as crankshafts, con-rods etc., which experience High Cycle Fatigue (HCF) where, typically, the number of loading cycles to failure N_f exceeds 10^5. The relationship of the material fatigue curve to a specific component feature and loading condition is via empirical factors and 'rules of thumb' to cater for cumulative damage, surface roughness, stress concentrations resulting from notches and geometric discontinuities, mean stresses and multiaxial stress states [1]. This approach was found to be inadequate when required to handle lightweight designs, where the operating stresses were likely to be in excess of the yield point of the material. Hence over the last three decades, with a major effort supported by the Society of Automotive Engineers in the USA [2], a methodology based on Strain-Life (ε-N) has been developed to model the elastic-plastic deformation of materials. These models are typically valid for Low Cycle Fatigue (LCF) applications ($10^3 < N_f < 10^5$). Much of the validation of these theories, however, has been carried out on laboratory specimens.

The ε-N approach suffers from serious limitations when applied to practical production components. The basic algorithms are inadequate when dealing with real surface conditions (roughness as found on 'as-cast' or 'as-forged' surfaces, residual stresses, etc.) and multiaxial stress states. The problem of multiaxiality is considered vital for lightweight design of solid parts (suspension linkages, engine parts, etc.) which experience multiaxial stress states as a result of combined loading or as a result of complex geometry which are unavoidable on practical components.

A further points merits discussion. The automotive industry is a 'volume' production industry with parts manufactured in large batches. There is a likelihood of significant part-to-part statistical variations in dimensional tolerances, surface roughness characteristics and residual stresses. Hence a 'reliability' oriented design approach, based on a probabilistic description of the design variables to predict the probability of survival for an entire batch is needed.

The objective of the investigation reported in this paper were therefore to validate some of the current fatigue life prediction approaches for dealing with the effect of notches and surface condition on mechanical parts in both Low Cycle Fatigue (LCF) and High Cycle Fatigue (HCF) regimes. Of particular interest was the comparison of the behaviour under constant and variable amplitude loading. Furthermore, since the part selected was manufactured in large batches, with a likelihood of significant part-to-part variation in dimensional tolerances, surface roughness and residual stresses, the validity checks on some probabilistic approaches, where these factors were treated as random variables, were also attempted.

TEST PROGRAMME

Component Test Programme

Component Description. The mechanical component selected for this investigation, as shown in Fig. 1a, forms part of a front suspension of a car. The component provides a typical example of a current production part which is manufactured in large batches. It essentially consists of a short, solid cantilever beam, attached at right angle to a thick-walled tube. This component, when mounted in the car, is free to swing freely on a pair of bearings embedded in the thick-walled tube section. The vertical wheel loads at the free end of the cantilever section are reacted by a suspension damper, mounted at the raised boss located about half way along the beam. The fillet radius at this boss represents a stress concentration feature because of the changes in geometry. In addition, this critical zone has a surface skin with 'as-forged' surface roughness characteristics. The investigation focused on the initiation of fatigue cracks at the root of this feature.

The component batch was hot die forged from a medium carbon steel, En15 (150m36), which was heat treated to 'R' condition. The composition of this material is indicated in Table 1 below.

Table 1 Chemical composition of En15R (in wt %).

C	Si	S	P	Mn	Ni	Cr	Mo
0.40	0.17	0.004	0.022	1.59	0.17	0.07	0.02

The material tensile properties were: $R_m = 830$ MPa, $R_{p0.5} = 580$ MPa and $E = 195$ GPa.

Component Test Set-up. The components were tested under bending with a servo-hydraulic actuator as shown in Fig. 1a. The component tests were performed under load control. Crack initiation in the fillet radius was sensed using an 'Alternating Current Potential Drop' (ACPD) system. The drop in demanded load signal correlated with the ACPD signal, hence the majority of the tests were monitored automatically by the test software. At failure, a typical fatigue crack was about 10 mm long on the surface and around 1 mm in depth.

The servo-hydraulic actuator control system could deliver either a constant amplitude sine wave loading history with adjustable frequency and amplitude, or a variable amplitude time history. The latter was based on the so called 'CARLOS-VERTICAL' [3] loading sequence for car suspensions. This standardised load sequence for car wheel suspension components has 136,084 loading cycles and represents a mission of 40,000 km on European roads. The sequence consists of variable amplitude cycles superimposed on a mean stress.

Component Surface Characterisation. Residual stresses (in the longitudinal direction) were measured on a batch of components using X-ray diffraction techniques. The measurement of surface roughness around the notch radius proved difficult. However, this was identical in character to the roughness of the straight surface close to this notch, hence these surface roughness measurements were used. The maximum peak-to-valley height, 'r_t' and 'r_a' which represents the arithmetic mean of the departures of the roughness profile from the mean line, were measured in the gauge length of about 10 mm using a Telysurf machine.

The components were fatigue tested in the 'as-received' condition with the critical site exhibiting 'as-forged' surface roughness. Some baseline tests were also conducted with the notch area hand-polished, using grade 1200 emery paper to remove the surface roughness.

Specimen Test Programme

Specimen Description. In order to facilitate a detailed characterisation of the En15R steel for its fatigue behaviour, special hour-glass-shaped, cylindrical specimens with parallel gauge sections were prepared, see Figs. 1b and 1c. The material used for these specimens was obtained from standard barstock (several batches) and also out of the actual production components (from longitudinal and transverse sections). In addition, a special batch was also hot die forged from the same material to obtain 'as-forged' surface condition. The heat treatment and the surface finish were practically identical to those for the production component.

Specimen Fatigue Tests. The specimens, both in the ground or in the 'as-forged' surface condition, were fatigue tested under strain control with fully reversed push-pull loading in a servo-hydraulic test machine (MTS type 810) for the baseline material characterisation in the LCF regime. However, a few tests were also performed under load control in the HCF regime for fully reversed stress (R=-1) and with a positive mean stress (R=0). Failure in the LCF regime was defined as 5% load drop-off which coincided with a surface circumferencial crack length of about 2 mm.

The same servo-hydraulic machine was also used for variable amplitude testing. For these tests, a modified version of the CARLOS-VERTICAL signal was used. The mean stress on the original signal was omitted together with minor cycles with amplitudes less than 25% of the largest amplitude in the sequence. The modified sequence had 31,289 cycles.

Specimen Surface Characterisation. A number of 'as-forged' specimens were randomly selected for measurements of the surface roughness which were conducted on a Telysurf machine. Two test lengths (top and bottom of the gauge section) of about 10 mm at 90° to the flash lines were traversed to obtain the centre line average roughness 'r_a' and the maximum peak-to-valley height 'r_t' measurements. Surface roughness measurements on the ground specimens were also obtained by the same method.

Surface residual stresses were measured using X-ray techniques. Both axial and circumferencial residual stresses were measured in the gauge section on the 'as-forged', ground and polished specimens.

TEST RESULTS AND ANALYSIS

Specimen Results

Constant Amplitude Results. A typical example of the fatigue response of En15R material in the ground condition under constant amplitude loading is shown in Fig. 2. There was little difference between various batches of En15R, provided the surface preparation was the same. A comparison of the ground condition with the 'as-forged' surface is shown in Fig. 3.

The surface roughness parameters 'r_a', and 'r_t' for the 'as-forged' surface were found to have average values of 8.3 µm and 54.4 µm respectively. These values, together with the tensile strength of the material were used to predict the surface finish factor, K_{sf}, using the ESDU curves [4]. This factor was used in Neuber's rule to predict the reduction in fatigue lives, as shown in Fig. 4a. There, in place of the fatigue notch factor K_f, the term K_t/K_{sf} was used, K_t being the geometric stress concentration factor; see below.

The residual stress measurements indicated an average compressive stress of about 272 MPa in the longitudinal direction. These figures were used in Morrow's equation [5]

$$\frac{\Delta \varepsilon}{2} = \frac{\sigma_f{'} - \sigma_m}{E} \left(2N_f\right)^b + \varepsilon_f{'} \left(2N_f\right)^c \tag{1}$$

to predict their effects on the fatigue performance of En15R, as shown in Fig. 4b.

Variable Amplitude Results. The variable amplitude test results obtained on specimens with ground and 'as-forged' surface finish are shown in Fig. 5. The predicted curves for the appropriate surface condition using Miner's summation and smooth specimen data (as shown in Fig. 2) are also included on the same plot.

Component Test Results

Constant Amplitude Results. The constant amplitude test results obtained under bending loads on the suspension arm are shown in Fig. 6. Results in the 'as-received' (as-forged) condition as well as with the critical site (where the cracks initiated) polished by hand to remove the surface roughness are included. The predicted curves based on the smooth specimen data mentioned above, stress concentration factor K_t, which was measured to be about 1.3, and the surface finish factor K_{sf} (estimated to be about 0.77 from ESDU curves) are also included in this graph.

Variable Amplitude Results. The variable amplitude test results (CARLOS-VERTICAL Sequence) are shown in Fig 7. The predictions based on smooth specimen data, on stress concentration factor ($K_t = 1.3$), on surface finish factor ($K_{sf} = 0.77$), and on the use of Miner's summation, are also included.

Probabilistic Analysis of Component Data

In this analysis, both the surface finish factor, Ksf, and the residual stress, σ_r, were assumed to be independent random variables. Experimental measurements carried out on a batch of components established the following statistical characteristics:

- normal density of the surface finish factor with mean value $E\{K_{sf}\} = 0.77$ and standard deviation $(\text{var}\{K_{sf}\})^{0.5} = 0.025$;
- normal density of the residual stress with mean value $E\{\sigma_r\} = -272$ MPa and a standard deviation $(\text{var}\{\sigma_r\})^{0.5} = 41.8$ MPa.

The above probabilistic inputs were used to calculate the probability of failure based on the Coffin-Manson equation,

$$\varepsilon_a = \left(1 - \frac{\sigma_r}{\sigma_u}\right)\frac{\sigma_f'}{E}\left(2N_f\right)^b + \varepsilon_f'\left(2N_f\right)^c \tag{2}$$

the hysteresis loop equation,

$$\varepsilon_a = \frac{\sigma_a}{E} + \left(\frac{\sigma_a}{K'}\right)^{\frac{1}{n'}} \tag{3}$$

and Neuber's rule,

$$\frac{\left(K_f S_a\right)^2}{E} = \sigma_a \varepsilon_a \tag{4}$$

The residual stress effects were predicted using the Goodman correction in the first term of the right-hand side of the Coffin-Manson equation, and the surface roughness effects were predicted substituting K_t/K_{sf} for K_f in Neuber's rule.

The "CPFI-Complete Probability Fast Integration" method [6], developed as part of this investigation and briefly described below, was used to predict the probability of failure as shown in Fig. 8a and 8b.

In the CPFI approach, the fatigue-life distribution $F_{\underline{N}}(N) = P\{\underline{N} \le N\}$ of a component, and hence the reliability $R(N) = P\{\underline{N} > N\} = 1 - F_{\underline{N}}(N)$, i.e., the probability that the random lifetime $\underline{N}$ does not fall below a given value N, is always expressed in a multiple-integral closed form. The life distribution is given by

$$F_{\underline{N}}(N) = P\{\underline{N} \le N\} = \int_{D_N} f_{\underline{\varepsilon}_a \underline{\sigma}_r}\left(\varepsilon_a, \sigma_r\right) d\varepsilon_a d\sigma_r \tag{5}$$

where $f_{\underline{\varepsilon}_a \underline{\sigma}_r}\left(\varepsilon_a, \sigma_r\right)$ is, in general, a joint probability density, and D_N denotes the failure region in the space of the random variable $\underline{\varepsilon}_a$ and $\underline{\sigma}_r$, i.e.,

$$-\infty < \sigma_r < \infty, \quad \left(1 - \frac{\sigma_r}{\sigma_u}\right)\frac{\sigma_f'}{E}\left(2N_f\right)^b + \varepsilon_f'\left(2N_f\right)^c \le \varepsilon_a < \infty \tag{6}$$

In contrast to approximate methods currently employed, in CPFI a rapid algorithm is developed which directly computes the right-hand-side of (5); it gives the exact solution, to any desired accuracy, for the entire ranges of fatigue-life and reliability. For further details on this method, see Bargmann et al. [6].

DISCUSSION

The analysis of the constant amplitude test data obtained on the 'as-forged' and on the ground finished specimens indicates that there is a significant reduction in fatigue strength under both LCF and HCF, resulting from the quality of surface finish. The data scatter in HCF does not show a clear distinction between the two surfaces. This is presumably due to the effects of surface roughness being counteracted by the presence of residual stresses. A prediction based on Goodman's correction shows a life improvement factor of about 1.3 due to the presence of the compressive residual stresses. However, this improvement is counteracted by the reduction in strength as a result of a reduced surface finish factor resulting from the surface roughness. These two effects cancel each other in the HCF and no clear distinction is apparent between the two sets of results.

However, in the LCF regime, it is known that the residual stresses will substantially relax away under plastic deformation, leaving only the strength reduction effects of surface roughness. The results demonstrate this effect very distinctly. The predictions based on the use of Neuber's rule correlate well with the measured data. Thus the surface roughness may be modelled as a network of notches.

Non-conservative estimates have been predicted for damage accumulation based on Miner's summation under variable amplitude history of long duration (number of reversals $> 10^5$). The agreement is within a factor of 2 which is in line with results reported in the literature. The reduction in strength is of a similar order under constant and variable amplitude loading.

The component lives, predicted from the simple specimen data, are within a factor of 2 as compared to measured lives. The predictions are on the conservative side. In part this may be due to an assumption of a plane stress condition in the notch whereas, in the practical situation, the notch condition was clearly plane strain. A very high degree of correlation between measured and predicted lives has been obtained under variable amplitude histories of long duration such as CARLOS-VERTICAL.

The probabilistic calculations have provided a practical method for determining a complete probability of failure versus life characteristic as opposed to the deterministic methods which only predict a single point at 50% probability of failure, for example. Although, the surface finish factor K_{sf} and the residual stresses were assumed to have normal densities in the above example, the application of CPFI is not restricted to this form of distribution; it is capable of handling any form of probability distribution for the input variables. While the CPFI method deals with physical uncertainties (such as variation in surface roughness and residual stresses) it does not handle the 'model' uncertainty; Miner's summation is hardly ever equal to 1.0. It is because of this model uncertainty why the predicted 50% probability line does not line up with the measured data.

CONCLUSIONS

The strength reduction effect due to surface roughness is accounted for by representing the surface as a collection of notches and making use of Neuber's rule. The geometric stress concentration factor, K_t

per unit surface finish factor, K_{sf}, the later estimated from both the maximum peak-to-valley height 'r_t', and the tensile strength of the material, is used in place of the notch factor K_f in Neuber's rule.

The strength reduction effects due to the surface roughness are similar under constant and variable amplitude loading. The same K_{sf} may be used for both types of loading.

Residual stress demonstrates a more pronounced effect under constant amplitude loading as compared to variable amplitude loading.

Cumulative damage under variable amplitude loading sequence of long duration on simple specimens is non-conservative by a factor of about 2 as compared to measured data. Possible reasons include (a) the effect of small cycles which, in principle, require non-linear summation, (b) the mode of relaxation of residual stresses.

The prediction of component lives based on specimen data is conservative by a factor of about 2. A possible reason for this effect may be due to an approximate assumption for the stress state.

ACKNOWLEDGEMENTS

The authors wish to acknowledge with thanks the support provided by Martin Hughes of European Gas Turbines and Martin Jones of Rover Group.

REFERENCES

1. Devlukia, J., "Software support tools for durability assessment and design of automotive components," Mathematics in the Automotive Industry, Ed: J. R. Smith, ISBN 0 19 8536607, 1992.
2. Wetzel, R. M., "Fatigue under complex loadings," SAE Publication 1977, Vol 4, 6.
3. Schutz, D., Klatschke, H., Steinhilber, H., Heuler, P., and Schutz, W., "CARLOS - Standardized load sequence for car wheel suspension components," LBF Report No. FB-191. Germany, 1990.
4. ESDU Publication 74027.- Endurance Data: Stress Concentration, Engineering Science Data Unit, London, 1982.
5. Morrow, J., Wetzel, R., and Topper, T., "Laboratory Simulation of Structural Fatigue Behaviour," ASTM, STP 462, 1970.
6. Bargmann, H., Rustenberg, I., and Devlukia, J., "Reliability of Metal Components in Fatigue: A Simple Algorithm for the Exact Solution," Fatigue and Fract. of Engng Mater. Struct., Vol. 17, No. 12, pp. 1445-1457, 1994.

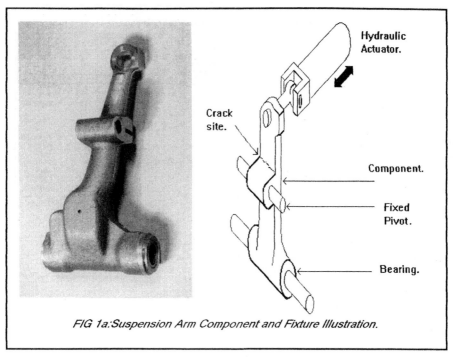

FIG 1a:Suspension Arm Component and Fixture Illustration.

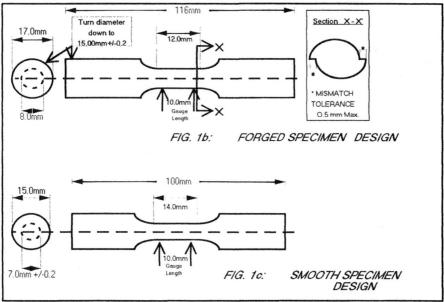

Fig. 1. Suspension arm component and test specimen.

J Devlukia et al.

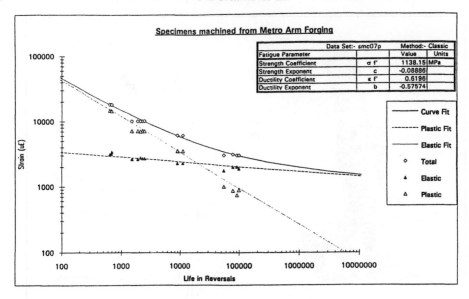

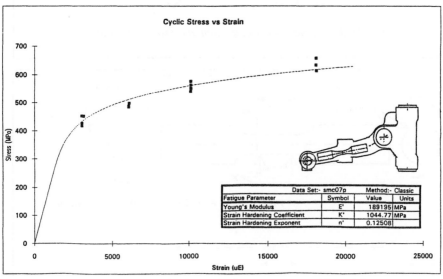

Fig. 2. LCF properties of EN 15R component material.

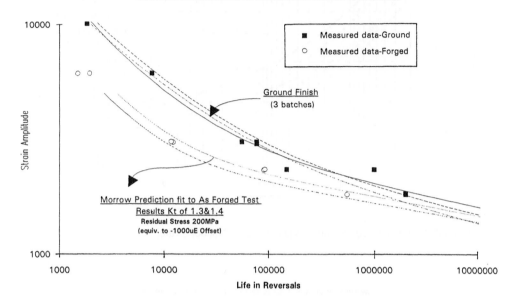

Fig. 3. Comparison of ground and "as-forged" finish on fatigue life of EN 15R specimens.

J Devlukia et al.

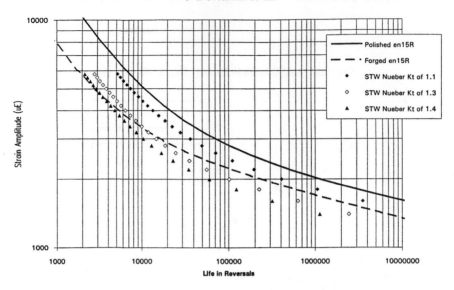

Fig. 4. (a) Prediction of surface roughness via the use of Nueber's rule.

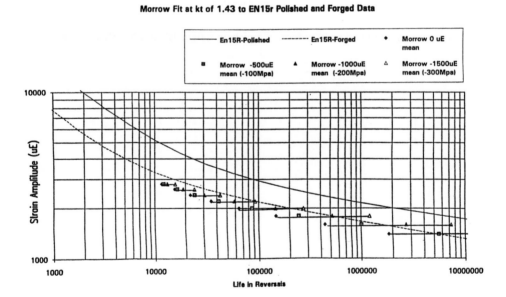

Fig. 4. (b) Effect of residual stresses predicted via Morrow correction.

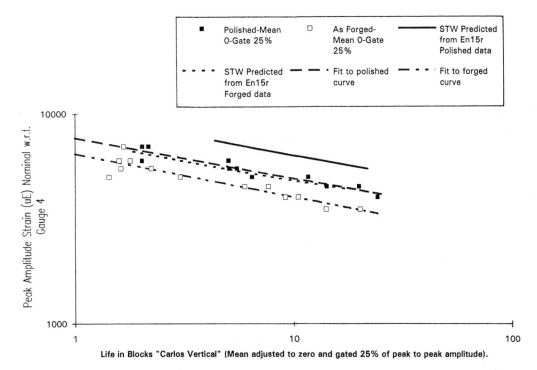

Fig. 5. Comparison of polished and "as-forged" surface effects on fatigue life of EN 15R specimens under variable amplitude loading.

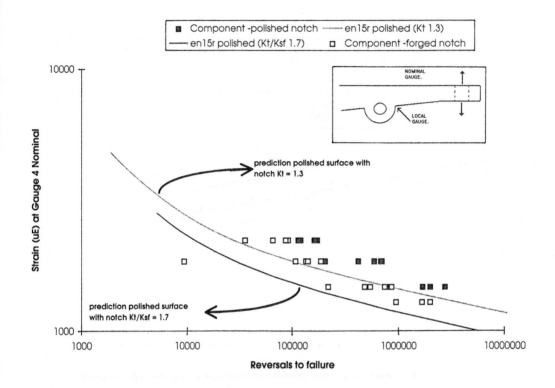

Fig. 6. Constant amplitude results for suspension arm component with prediction using Neuber at K_t of 1.3 and K_t/K_{sf} 1.7 using EN 15R polished material data.

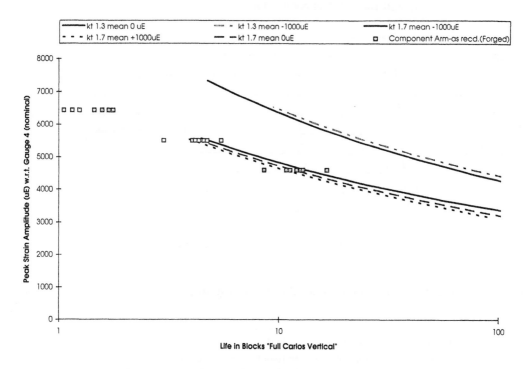

Fig. 7. Measured and predicted fatigue lives on suspension arm under variable amplitude loading.

**FATIGUE LIFE DISTRIBUTION FOR FORGED SUSPENSION ARM SUBJECTED
TO NOMINAL STRESS AMPLITUDE.**

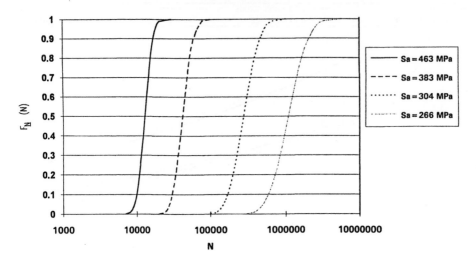

Fig. 8. (a) Probability of failure v/s lifetime at different nominal stresses.

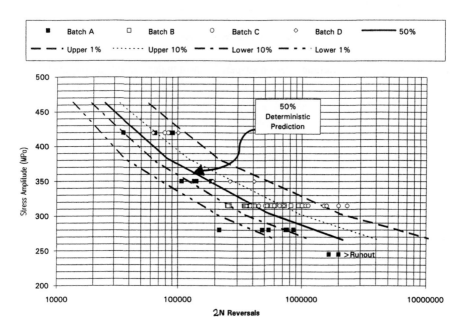

Fig. 8. (b) Probabilistic lifetime prediction based on the CPFI method (K_{sf} mean = 0.77, Sd = 0.025, residual stress mean = -272, Sd = 41.8 Mpa).

A FATIGUE LIFE ASSESSMENT METHODOLOGY FOR AUTOMOTIVE COMPONENTS.

Anne-Sophie Béranger, Jean-Yves Bérard, Jean-François Vittori
RENAULT, 860 quai de Stalingrad
F-92109 Boulogne - Billancourt cedex, FRANCE

ABSTRACT

In the automotive industry, most of the components are subjected in service to fatigue loading which may result in failures. In order to reduce design lead times and to assure a high reliability level of the parts, general procedures for durability assessment are developed. In this framework, this paper aims to present a fatigue life assessment methodology. Various factors are involved in the analysis : material fatigue data, multiaxial fatigue criteria, finite element method (FEM) calculations. A validation of this procedure was conducted via data generated from tests on a real current production component, namely a RENAULT Safrane suspension arm which was die cast from spheroidal graphite (SG) cast iron.

KEYWORDS

Fatigue criteria, Life prediction, SG cast iron

INTRODUCTION

Numerous studies on the fatigue behaviour have been carried out since the 50's. Parameters such as chemical composition, heat treatments and surface finish have been studied. The use of these data for component design is a difficult task for several reasons. The choice of a multiaxial fatigue criterion is of first order. Von Mises equivalent stress is still of common use, though it is well known that it may lead to unsafe results.

The demand for lighter vehicles, reduced design lead time and a good reliability is always increasing. Therefore, it is vital to develop improved modelling tools for fatigue assessment. Such tools will enable the designers :
   (1) To operate the materials at higher stress levels without increasing the mass of the components.
   (2) To reduce the number of design-testing iterations.
   (3) To assure the required reliability level.

In this framework, this paper aims to present a study of the fatigue behaviour of a spheroïdal graphite (SG) cast iron safety component, at room temperature. The basic methodology for fatigue assessment adopted is general. It can be extended to a variety of components subjected in service to fatigue. This methodology includes :
   (i) Extensive investigation of the fatigue behaviour of the material.
   (ii) Calculations of the stress state of the component via finite element method (FEM).
   (iii) Fatigue assessment by mean of a multiaxial criterion.
In this paper, a validation of the procedure will be presented, via fatigue tests on a real component.

17

TEST COMPONENT AND MATERIAL

The Safrane suspension arm

The scope of this study is to investigate the high cycle fatigue behaviour of an automotive suspension linkage, namely a front suspension arm from the RENAULT Safrane. Figure 1 shows the part and its installation on the car.

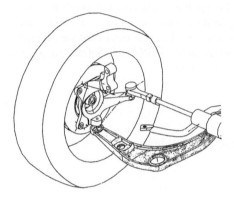

Fig. 1. Illustration of the suspension arm and its installation on the car.

This part provides a typical exemple of a current production part manufactured in large batches via die casting of ferritic cast iron. This component exhibits surface roughness due to the casting process and residual stresses as the result of the shot peening operation (to remove the sand from the mould and the weight feeds).

The test material

Using relevant material data is a very important point for reliable fatigue life assessment. This is the reason why an extensive test program was conducted from especially prepared specimens from the same material (SG grade with pearlite content between 0 and 25% in agreement with the specifications for component casting). These test data consisted of :

(1) Monotonic tension.
(2) High cycle fatigue behaviour.
(3) Low cycle fatigue deformation response.

The tensile mechanical properties of the machined material are the following :

Table 1. Mechanical properties of the material.

E	170 GPa
yield stress $\sigma_{y0,2}$	360 MPa
U.T.S.	490 MPa
elongation	9 %

Specimens were machined and tested in high cycle fatigue. The results were compared to :

(i) those for as-cast specimens : to take the irregular surface roughness produced by the sand mould into consideration.
(ii) those for shot-peened specimens : to take the compressive residual stresses into consideration.

Additionally, various types of loads (torsion, tension-compression) were tested, and various stress ratios were used, so as to account for the in-service loads. Finally, 25 specimens were tested in each case in order to estimate the scatter. The results are plotted on Fig. 2. for a stress ratio R = - 1 in tension-compression. Statistical analysis of these results were undertaken using the ESOPE code [1]. The lines represent the Wöhler curves with a failure probability of 50%.

It can be seen that the rough surface of the as-cast condition, only marginally reduces the fatigue limit : it is only 10% lower than the limit obtained on machined-surface specimens. The shot-peening on the as-cast surface provides a substantial improvement, around 30%, of the fatigue limit, due to the residual compressive stresses induced by the process. These results are in agreement with previously published data on cast-iron [2] [3].

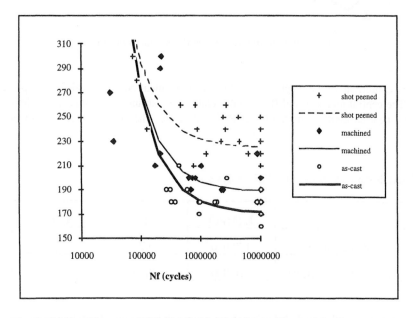

Fig. 2. Wöhler diagram of GS 52 cast iron relative to the three surface conditions
with statistical analysis curves (tension - compression R = -1).

F.E. CALCULATIONS

The fatigue life assessment methodolgy was developed in order to be integrated to the current design practises which are based on F.E. analysis. F.E. calculations were undertaken on the test part. A model with shell elements was used. The fatigue test performed on the part were simulated in the following way : The loading is in the plane of the suspension arm as a simulation of braking and acceleration (Fx). It is applied on the central node E (Fig. 3).

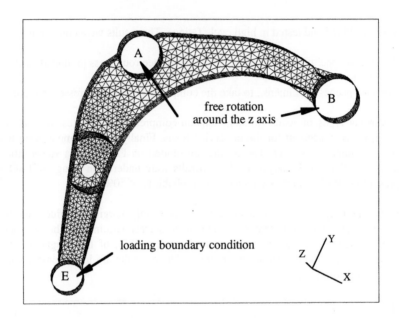

Fig. 3. Shell elements mesh and boundary conditions

The boundary conditions in the neighbourhood of the attachment points A and B are representative of the test rig. The central nodes A and B can only rotate freely around the z axis. The calculations were performed in linear elasticity using the NASTRAN F.E. code. The material parameters were as shown in table 2.

Table 2. Materials parameters used for F.E. analysis.

	E (MPa)	n	r (kg/dm^3)
SG cast iron	170 000	0,29	7,2
Steel (bores A, B, E)	210 000	0,29	7,8

The suspension arm was also analysed using strain gauges. It was shown from these results that there is a good agreement between the calculations and the strain measurements : the maximum difference is around 10%. The model provides a particularly good precision in the areas where the cracks occured during the experimental tests.

Figure 4 shows the Von Mises equivalent stress generated by the simulation of braking under a loading condition Fx = 10 000 N. In this case the most highly stressed region is located on the outer rim, close to point A (arrow). The maximum Von Mises equivalent stress is about 190 MPa. It was shown that this region is in tension when looking at the maximum principal stress.

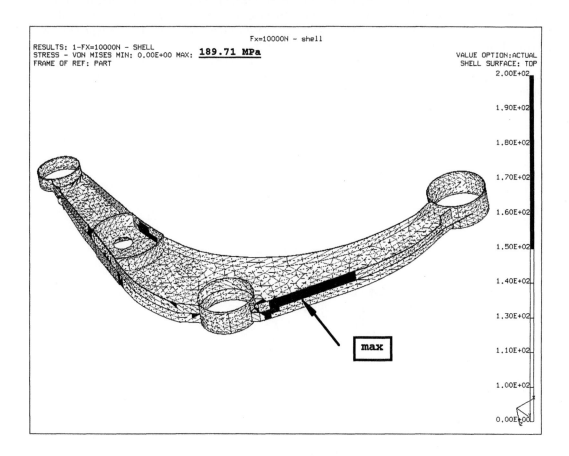

Fig. 4. Stress generated by the simulation of braking (Fx = 10000 N).

As far as the acceleration loading case (Fx = - 10 000 N) is considered, a similar value of the equivalent stress is found, due to the hypothesis of linear elastic behaviour, but in this case the maximum principal stress is compressive.

Although the same maximum equivalent Von Mises stress was found, these two cases exhibit opposite principal stresses leading to different fatigue behaviour.

COMPONENT TESTING

Testing conditions

The objective of this test programme was to evaluate the fatigue life on the component under study. The actual endurance limit was determined at $2 \cdot 10^6$ cycles.

Figure 5 presents the actual test rig set up used to perform the fatigue tests on the component. The tests were performed under load control with constant amplitude. The frequency was 10 Hz. The load was imposed along the X axis direction with R = - 0,5. This load ratio was chosen since it is representative of typical braking / acceleration cycles measured on vehicle.

Fractographic analysis was conducted on the tested parts. Fatigue crack initiated at the surface on the outer rim of the part near the point A. These observations are consistent with the location of the highest tension principal stress from the F.E. calculations.

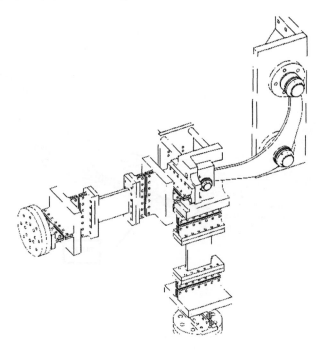

Fig. 5. The test rig

Results

In the fig. 6, cycles to failure were plotted versus the maximum load. The curve was obtained through a statistical analysis using the ESOPE code with a failure probability of 50%.

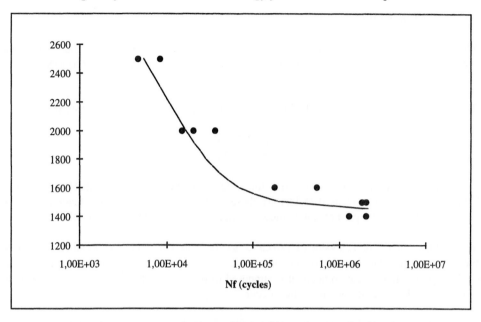

Fig. 6. Wöhler curve on the test component subjected to Fx loading at R = - 0,5
(load vs cycles).

From these experimental results, the fatigue limit of the suspension arm under uniaxial loading condition and with a stress ratio R = - 0,5 is about 1 450 daN. From the above linear elastic calculations, at this load level, the maximum Von Mises stress is 190 x 1,45 = 300 MPa which is below the yield stress, and the corresponding tension maximum principal stress is 285 MPa. The conclusion that a linear elastic simulation is appropppriate and relevant for our study.

FATIGUE ASSESSMENT

One goal of this work was to verify the pertinence of the multiaxial fatigue model proposed by Dang Van [4].

$$\tau(t) + \alpha\, p(t) - \tau o > 0 \qquad\qquad (1)$$

Using relevant material data, it is expected to predict the actual fatigue limit on the component with F.E. analysis. In this section, the fatigue life prediction code developed at RENAULT will be presented. Fatigue life prediction of the component will be performed and analysed.

Fatigue analysis code

A fatigue life prediction code has been developed and used at RENAULT for almost 10 years [5]. Several multiaxial models such as Crossland [6], Sines [7] [8], Dang Van [4] [9] and Deperrois [10] are programmed : for Dang Van criterion, both the critical plane and the hypersphere procedures [9] are implemented. This latter procedure is most of the time preferred due to shorter computation time.

Fatigue life prediction results.

Fatigue life prediction has been performed on the test component. The result of the static linear elastic F.E. calculation is used to generate the load cycle. This one is +1 450 daN ; 0 ; -720 daN; 0 ; +1 450 daN.

The Dang Van fatigue model was used with a material damage line, whose coefficients are :

$$\tau_0 = 217 \text{ MPa}$$
$$\alpha = -1, 38$$

They correspond to the shot-peened material, for a life of 10^7 cycles, with a probability of failure of 50%. Several stress ratios are used to identify these coefficients. Therefore any loading cycle, with various stress ratios, can be predicted with these coefficients using the Dan Van model.

The result of the calculation is a value of a so called safety factor (Sf). This factor can be defined as the normalised smallest distance of the loading path in the (τ ; p) Dang Van diagram to the damage line. When Sf is greater than 1 the component is supposed to be safe. On the contruary when Sf of some elements become less than unity, failure may occur.

Isovalues of the safety factors are presented on Fig. 7 for the overmentionned fatigue test on the component.

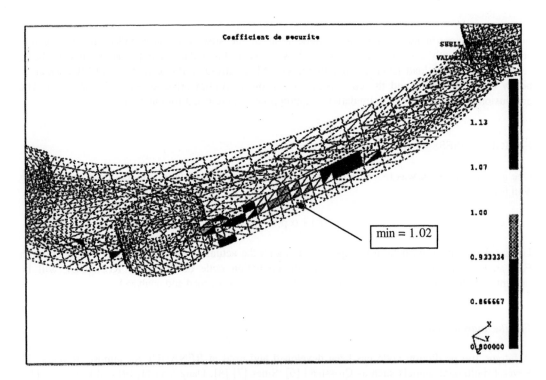

Fig. 7. Isovalues of the safety factors for the suspension arm tested in fatigue.

The minimum value of the safety factor on the part is located in the area where the failure occured on the test rig. Its value is 1,02. This result exhibits a very good correlation between the experiment and the fatigue life prediction since the theoretical value of 1 and the actual safety factor are very close.

Several damage lines representative of the various surface finish have been tested for the predicition of the safety factors associated to the fatigue life. It was shown that the safety factors obtained are highly dependent on the material data. The best results were derived from material data representative of the exact surface finish of the part, namely shot-peened.

CONCLUSION

A fatigue life assessment methodology for automotive components has been presented.

This one is based on a static F.E. linear alstic anlaysis of the component followed by a fatigue analysis using the Dang Van fatigue life prediction model. On the basis of material fatigue data, the safety factor can be estimated for every element of the model.

To validate this methodology, fatigue tests on real components have been performed. For one loading case, an endurance limit on the component was evaluated. Using relevant material fatigue data, safety factor on the suspension arm were computed. A very good correlation was found between the predicition of the model and the actual tests results on the component.

REFERENCES

1. Bastenaire, F., (1971). Revue Française de Mécanique **37**, pp. 25-36.
2. Parent-Simonin, S., Fargues, J. and Flavenot, JF. (1988). *Traitement Thermique* **221**, 17.
3. Starkey, MS., Irving, PE. (1982) *Int. J. Fatigue*, 129
4. Dang Van, K., (1973). *Sur la résistance à la fatigue des métaux*, Sciences et Techniques de l'Armement, Mémorial de l'Artillerie Française, 3ème fascicule.
5. Diboine, A., (1994). In : *Proceedings of the 4th Multiaxial Fatigue Conference*, ESIS , Paris, pp. 93-109
6. Crossland, B., (1956). In : *Proceedings of the International Conference on Fatigue of Metals*, Institution of Mechanical Engineers, London.
7. Sines, G., (1959). *Metal Fatigue*, G. Sines and J.L. Waisman (Eds). McGraw Hill, pp. 145-169
8. Sines, G., (1955). *Failure of materials under combined repeated stresses with superimposed static stresses*, Techn. note 3495, National Advisory Comittee for Aeronautics, Washington DC, 69.
9. Dang Van, K., Griveau, B. and Message, O., (1989). In : *Biaxial and multiaxial fatigue, EGF3*, Mechanical engineering Publications, London, pp. 479-496.
10. Deperrois, A., (1991). *Sur le Calcul de Limite d'endurance des Aciers*, PhD thesis, Ecole Polytechnique, France.

FATIGUE STRENGTH PREDICTION FOR GREY CAST IRON COMPONENTS OF COMPLEX GEOMETRY

D.H.Allen, P.M.Hughes

GEC ALSTHOM Power Generation, Mechanical Engineering Centre,
Cambridge Road, Whetstone, Leicester, LE8 6LH

ABSTRACT

A methodology for fatigue design assessments of complex cast iron components is presented. The approach is based on the concept that the fatigue strength of a component is dependent on whether or not a crack will grow out of a very localised high stress field. The need to define a nominal stress or Kt factor is avoided, and so allows the direct use of finite element analysis results.

Comparisons between predicted and actual test specimen fatigue strengths appear quite favourable. The need for further work is identified in order to: (a) gain confidence in the procedures, and (b) generalise it to other materials.

KEYWORDS

crack growth, fatigue, grey cast iron, finite element analysis, notches.

NOMENCLATURE

a_0 El Haddad short crack parameter [1]; the crack length at which the value of stress range needed to create a ΔK value equal to ΔK_{th} is the same as the fatigue limit of the uncracked material.

a_1 Size below which cracks have no effect on the fatigue limit.

F_E K correction factor for ellipticity

F_G K correction factor for non-uniform stress field

F_S K correction factor for free surface

K_f Ratio of fatigue strengths: Smooth specimens/Notched specimens

K_t Elastic stress concentration factor

$\Delta K_{app,th}$ Apparent short crack growth threshold (crack size dependent)

INTRODUCTION

Fatigue cracking in engineering components is generally found to initiate at regions of stress concentration. Modern stress analysis techniques such as the finite element method provide the capability to accurately determine stress and strain fields around such features. However, when it comes to predicting the fatigue strength, it becomes apparent that existing fatigue design methods which utilise K_t, K_f and uniaxial nominal stress can be inadequate and have become out of step with current stress analysis methods. In a typical engineering component the stresses will vary continuously, and a corresponding nominal stress will be difficult if not impossible to define.

What is required are methods based on the complete stress and strain field local to the stress concentration. In other words, a method based on what is actually controlling the fatigue cracking process. Smith and Miller [2] identified two regimes for notched fatigue specimens: (i) below a certain K_t the fatigue strength depends on the local surface stress level, (ii) above this, notches behave the same as sharp cracks of the same depth as the notch. In a typical engineering component however, it is often difficult, if not impossible to define a notch depth at a stress concentration. The estimation of the value of a stress intensity factor which is applicable to the stress concentration in question will therefore also be difficult.

This paper describes an attempt to develop a method to predict the high cycle fatigue strength of engineering components of complex geometry. Specifically, engine components in grey cast iron for which detailed finite element analysis results are available. The method has been applied to three different specimen geometries at different mean stress levels, and the results have been compared with the experimentally measured values.

The work which was carried out involved a combination of experimental testing of notched fatigue specimens, and theoretical analysis based on the finite element method. The experimental procedure is described below, followed by a section describing the theoretical model which is proposed. The results are then presented and discussed.

EXPERIMENTAL PROCEDURE

The material used for the fatigue tests was grade 17 (260 MPa) flake graphite cast iron, obtained from 50mm diameter cast test bars. The graphite flake structure was Type A in a pearlitic matrix. High cycle fatigue tests were carried out at ambient temperature on both plain and notched specimen geometries, using an Amsler vibrophore resonance machine at a frequency of 100 Hz and various mean stress levels.

Three notch geometries were considered; two circumferentially grooved specimens as shown in Figs. 1(a) and 1(b), and a specimen with a shoulder fillet shown in Fig. 1(c). The latter was intended to be more representative of actual component details, while retaining the simplicity of a compact axial test piece. 18, 12 and 23 tests were performed on each of the notched geometries respectively. Testing was continued on each specimen until complete fracture or an endurance of 3×10^7 cycles, whichever was sooner. Fatigue crack growth threshold (ΔK_{th}) tests were also carried out using compact tension specimens at several R-ratios.

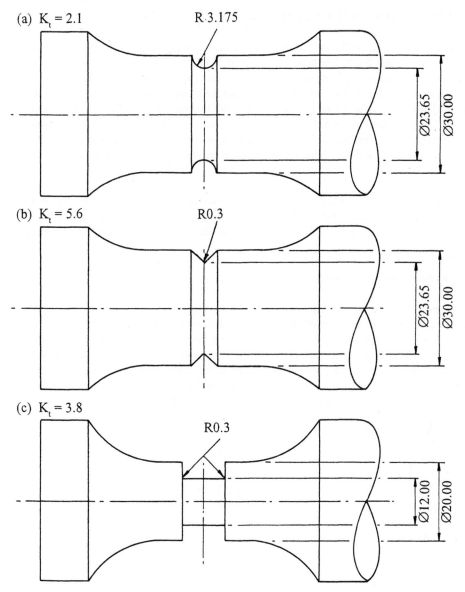

Fig. 1 Test specimen geometries

D. H. Allen and P. M. Hughes

DESCRIPTION OF CRACK GROWTH BASED PREDICTIVE MODEL

Under cyclic loading, particularly at high tensile stresses, deformation of the matrix material occurs and critically oriented graphite flakes crack or debond from the matrix to form crack-like features. This has been observed by Fash et al [3] and Jianchun [4] using surface replication techniques. Many cracks are observed to form over the surface, and these appear to grow by coalescence. Castillo and Baker [5] have studied grey cast iron in fatigue using plain specimens. They concluded from their work that the fatigue limit is dominated by the graphite eutectic cell size and is insensitive to the microstructure of the matrix, and a fracture mechanics model can be used to predict the fatigue limit based on an initiating defect which is determined by the eutectic cell size, and the threshold stress intensity for propagation.

Of interest in this work is whether or not the cracks will grow to failure. The Kitagawa-Takahashi plot [6] has been shown to describe the effect of short crack size on the threshold stress level [1]. This work has sought to extend the use of this model to the complex stress fields around stress concentrations.

A Model for a Crack in a Notch

It is assumed that the localised high surface stresses which are experienced at the notch root (see Fig. 2) will cause many of the graphite flakes at the surface to crack open and form an array of surface cracks. It is further assumed that if the crack driving force under cyclic loading exceeds the threshold condition for growth to occur, then each of these will grow in size until a point is reached when many will coalesce. The resulting continuous surface crack will experience a significantly higher crack driving force, the critical period is therefore likely to be just prior to the formation of a continuous surface crack by coalescence. Let a_c denote the crack depth at this point.

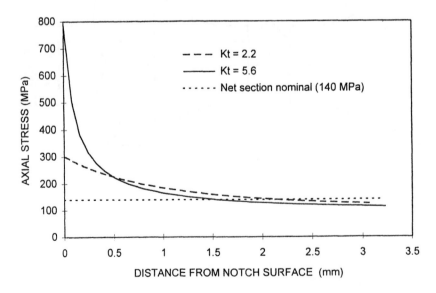

Fig. 2   Comparison of elastic stress distributions below the surface of two of the geometries

The proposed assessment procedure for a component subjected to a given cyclic load is to determine whether the ΔK value for a growing crack is exceeded by the apparent short crack threshold ΔK_{th} before the crack geometry transforms into a continuous surface crack by coalescence (see Fig. 3):

a)   For several crack sizes up to a_c
     - Estimate ΔK due to the applied load (LEFM assumed)
     - Estimate the local crack growth threshold value.

b)   Define the fatigue strength reserve factor as the maximum value of:

$$RF = \frac{\Delta K_{th}}{\Delta K} \qquad (1)$$

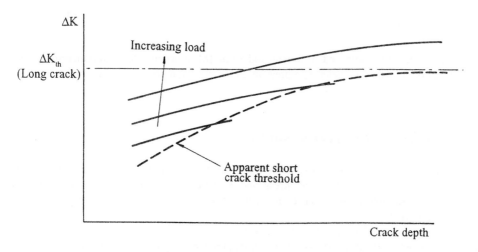

Fig. 3   *Schematic of ΔK versus crack length compared with crack growth threshold*

Estimation of ΔK for a Notch Root Surface Crack

The solution for the stress intensity factor of a surface crack subjected to non-uniform distributed stresses is given in the form as described by Albrecht and Yamada [7]:

$$K = F_E \cdot F_S \cdot F_G \cdot \sigma \sqrt{\pi a} \qquad (2)$$

In which F_E is a correction factor for ellipticity, and is equal to 0.65 for a semi-circular crack. F_S is a free surface correction factor (=1.12), and F_G is a crack size dependent factor to account for the non-uniform stress field.

Comparisons with three-dimensional finite element crack analyses of these notch geometries indicated that an approximate solution may be obtained using the simple expression:

$$F_G = \frac{Uncracked\ body\ stress\ at\ deepest\ point}{Uncracked\ body\ stress\ at\ surface} \tag{3}$$

The effective R-ratio at a point on the crack front will be influenced by plasticity during initial load-up. This effect was estimated at the deepest point on the crack front using a Neuber approximation, taking into account the multiaxial stress state. Several three-dimensional finite element crack analyses with superimposed residual stress fields have indicated that this approximation is not unreasonable. However a more complete validation of this procedure is still required.

Estimation of Crack Growth Threshold

Crack sizes of interest in this work are relatively short. By reformulating the conventional Kitagawa-Takahashi plot [6] of short crack threshold stress with crack size, in terms of an apparent short crack threshold (ΔKth), the following equation can be used:

$$\Delta K_{app,th} = \Delta K_{th} \left[\frac{a}{a - a_1 + a_0} \right]^{1/2} \tag{4}$$

This is based on the equation proposed by Lukas and Kunz [8] to account for non-damaging cracks below a size given by a_1. The parameter a_0 represents the intersection point of the plain fatigue limit with the long crack threshold (ΔK_{th}) line.

EXPERIMENTAL AND PREDICTIVE RESULTS

For each of the three specimen geometries which were tested, predictions of fatigue strength were made. In each case, the use of a nominal stress and K_t value was avoided to ensure that only information typically available in an actual component assessment was used. For comparison, two approaches have been used:
(a)   The proposed short crack growth based approach
(b)   The conventional local surface stress-strain based approach using the plain specimen
        fatigue strength.

Two different graphite flake sizes have been considered for approach (a) to illustrate sensitivity and likely bounds on the solution. Jianchun [4] estimated an average flake size of 0.12mm for this material. Results are also presented for an assumed flake size of 0.3mm. In each case, the value of a_c is taken as double these values, with the assumption that in these dense arrays of graphite flakes, if each flake grows, they will coalesce at twice their original size.

Figures 4 and 5 show the results for the circumferentially grooved specimens (K_t = 2.1 and 5.6) in terms of nominal net section stresses. Fig. 6 shows the results for the specimen with the shoulder fillet (K_t = 3.8) in terms of applied load.

DISCUSSION

The short crack growth based predictions provide more realistic estimates of the fatigue strengths than the predictions based on the local surface stresses. The latter are clearly highly pessimistic for the sharper notches. When based on the estimated average flake size, the short crack predictions lie within all the failure points for this data and are therefore conservative. The flake size of 0.3mm is likely to be an upper bound, and apart from at high mean stresses, lies at the upper edge of the run-out data points.

At high mean stresses, plasticity due to the initial loading will be significant, particularly close to the notch surface. The present predictions have attempted to account for the effects of sub-surface plasticity on the effective R-ratio at the deepest point on a crack. At the surface however, it can be speculated that extensive yielding and the resulting residual stress field may cause premature crack closure. This type of effect has not been included in the model. At high mean stresses the present model significantly underpredicts the strength.

A major assumption made in the present work has been that because the notch surface stresses will be very high, many of the graphite flakes at the surface will fracture to form a dense array of cracks, and that these will coalesce to form a continuous surface crack once they have grown to approximately twice the flake size. For flake iron, this needs to be tested on other grades where the flake size differs significantly from the present grade.

Flake graphite cast iron is peculiar in that a large number of cracks are formed. In other materials, such as spheroidal graphite cast iron, the crack density at the notch root is likely to be significantly different. Therefore, whether crack growth is considered in isolation, or whether coalescence is an important factor would require careful consideration for this type of mechanistic model.

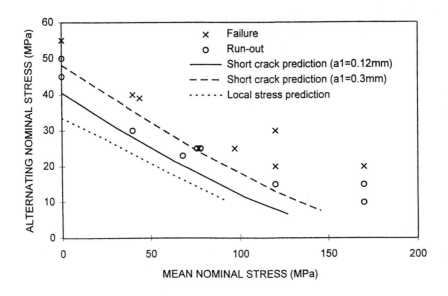

Fig. 4 Comparison of predicted and measured fatigue strengths for a notched specimen (K_t 2.1)

D. H. Allen and P. M. Hughes

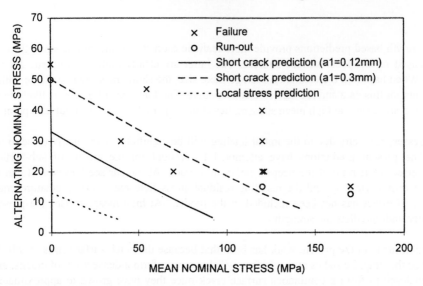

Fig. 5 *Comparison of predicted and measured fatigue strengths*
     for a notched specimen (K₁ = 5.6)

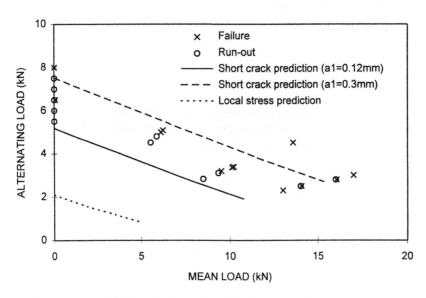

Fig. 6 *Comparison of predicted and measured fatigue strengths*
     for a notched specimen (K₁ = 3.8)

CONCLUSIONS

A crack growth based methodology has been proposed for predicting the fatigue strength of engineering components. It addresses the problem of carrying out fatigue assessments directly from finite element analysis results for which a nominal stress and K_t cannot be defined.

Comparisons have been made between predictions and test results for three test specimen geometries in grey cast iron. For estimated average graphite flake sizes, the predicted fatigue strengths are conservative and lie below all failure data points.

The proposed methodology provides significantly improved estimates of fatigue strength compared with predictions based on local surface stresses. Further work is required, however, to generalise it to other materials.

REFERENCES

1. El Haddad, M.H., Topper, T.H., Smith, K.N. (1979). Engineering Fracture Mechanics, **11**, pp.573-584.
2. Smith, R.A., Miller, K.J. (1978). International Journal of Mechanical Science, **20**, pp.201-206.
3. Fash, J.W.,Socie, D.F., Russell, E.S. (1981). In: Proceedings of Fatigue '81, Warwick University, SAE, pp. 40-51.
4. Jianchun, L., (1993). PhD Thesis, Mechanical and Manufacturing Engineering Department, University of Dublin, Trinity College.
5. Castillo, R.N., Baker, T.J. (1984). In: Advances in Fracture Research, New Dehli, India, Vol. 3-Q, pp.2057-2064.
6. Kitagawa, H., Takahashi, S. (1976). In: Proc. 2nd Int. Conf. Mechanical Behaviour of Materials, Boston, American Society of Metals, p.627.
7. Albrecht, P., Yamada, K. (1977) Journal of the Structural Division, Proceedings of the American Society of Civil Engineers, **103**, No ST2, pp. 377-389.
8. Lukas, P., Kunz, L., (1989). Fatigue and Fracture of Engineering Materials and Structures, **12**,(3), pp. 175-186.

DEFECT SENSITIVITY IN NODULAR CAST IRON FOR SAFETY CRITICAL COMPONENTS

KENNETH HAMBERG, BENGT JOHANNESSON AND ANDERS ROBERTSON
AB Volvo, Technological Development
Göteborg, Sweden

ABSTRACT

In this investigation the defect sensitivity of nodular cast iron in fatigue loading has been analysed. All specimens have been taken from components with an as cast surface marred by clearly visible surface defects. The study and the calculations made are based on fracture mechanics, and constant amplitude loading. A computer program has been designed to calculate the fatigue life for the specimens based on Paris law. The calculations show good agreement with actual cycles to failure. The results show that visible surface defects with a magnitude of a=0.8 mm (depth) and c=2.3 mm (half width), which often lead to rejection of a component, do not lower the fatigue life noticeably.

KEYWORDS

Nodular cast iron, defects, life time prediction, fatigue testing, S-N-curve,

INTRODUCTION

Nodular cast iron is a material that has found a increasing number of applications in the automotive industry during the last decade. It has a static strength comparable to cast steels and a greater fatigue strength and ductility than grey irons. Castability and machinability is good, and all these properties makes it an economic alternative for medium stressed components and for safety critical applications. A reduction of 30% or more in component cost can be made when nodular iron is substituted for cast or forged steel [1]. Nodular iron is however a material that contains different kinds of defects, such as inclusions, dross, surface defects, slag stringers and micro shrinkage pores. It has been found that cracks will initiate and propagate from these defects. The degree to which these different defects will lower a components service life is not fully investigated [2, 3]. A rule of thumb is that the rough cast surface lowers the fatigue performance to the same level independently of the grade of the material.

The present work is a study of the lifetime for test specimens in nodular cast iron with an as cast surface marred by surface defects. The analysis and calculations are based on fracture mechanics.

EXPERIMENTS

Material

Specimens have been produced from two different components which had slightly different matrix and will be called material A and B. Before any testing was made, the specimens were divided into two different groups. Group OK declared free from surface defects, group D for specimens declared defect from a visual inspection point of view. To be declared defect a specimen had to contain a surface defect in the size of approximately 2-3 mm. All specimens are tested with a "real" cast surface that is shot blasted.

Chemical analysis

The chemical analysis for material A and B are shown in Table 1.

Table 1. Chemical analysis of material A and B.

Mtrl	C	Si	Mn	P	S	Cr	Ni	Mo	Cu
A	3,65	2,59	0,36	0,010	0,005	0,06	0,08	0,01	0,19
B	3,49	2,35	0,32	0,013	0,012	0,06	0,07	0,02	0,09

Matrix and microstructure

There exists a pearlite rim at the surface of the component with a depth of about 0.07-0.15 mm. Microstructural constituents of the bulk material are given in Table 2.

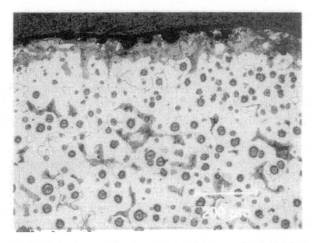

Figure 1. Optical micrograph of material B (Neg nr 78232 (100X)).

Table 2. Microstructural data for material A and B.

	Material A	Material B
Ferrite (%)	64	53
Pearlite (%)	22	27
Nodule count/mm^2	200	220
Nodule ø (µm)	27	20
Ferritic grain size (µm)	26	30

Tensile test and residual stress measurements

Tensile tests have been carried out on each material and a mean value from at least three specimens is presented in Table 3.

Table 3. Mechanical properties for material A and B.

	Material A	Material B
R_m (MPa)	512±11	476±13
$R_{p0,2}$ (MPa)	337±14	310±10
A (%)	12.0±1.0	14.3±3,3

An analysis was made on specimen OK2$_A$ and D10$_B$ to get an apprehension on the residual stress situation on the as cast surface The results are shown in Table 4.

Table 4. Residual stress on specimen OK2$_A$ and D10$_B$ at the as cast surface.

	OK2$_A$	D10$_B$
Normal stress (MPa)	-345±14	-250±14
Shear stress (MPa)	0	20±3

Fatigue crack simulation

To calculate the fatigue life for the specimens a computer program has been designed. The program is based on stress intensity factor calculations made on the defects that initiate the fractures and on Paris law. The stress intensity factor equations used are derived by Newman and Raju [4], who have applied the finite-element method to produce an approximate analytical method. It has been noticed that even when $R>0$ a crack might be partially closed during part of the loading cycle [5]. As a result the applied stress intensity has to be adjusted to an effective one, $\Delta K_{eff}=K_{max}-K_{op}$. At low ΔK values the R ratio becomes increasingly important for the crack propagation. In this crack growth regime, different crack closure mechanisms have been identified to influence K_{op} levels and associated ΔK_{eff} values. The values for nodular iron found in the literature on the effective ΔK_{th}, $R=0$, range from 6 to 8.5 MPa√m for long cracks in bending, see Fig 2. The lower values represents a pearlitic matrix and higher values a ferritic [6,7]. The sensitivity of ΔK_{th} to R ratio for a given material depends on the observed level of crack closure.

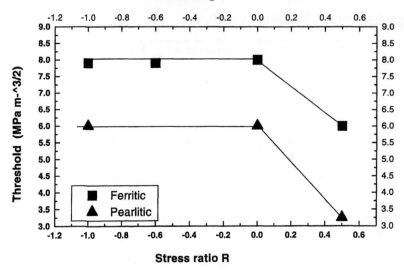

Figure 2. Correlation between effective threshold stress intensity factor range DK_{th} and stress ratio, R for long cracks [7].

The wide scatter of the material constants C and m complicates the fatigue life calculations [7,8,9]. All calculations in this study are made with $C=3.1*10^{-10}$ mm/cycle and $m=3.9$. These values are taken from [10], representing a ferritic-pearlitic matrix.

Short Cracks

If a fatigue crack initiates and propagates from a small defect or a micropore, the problem of short cracks come into focus. Short cracks ($a<0.5$ mm) tend to propagate faster than would be inferred from crack propagation data determined from long cracks ($a>0.5$ mm). This has been the conclusion from growth experiments carried out on cracks that initiated at "natural" sites such as inclusions and pores [11,12,13]. It was observed that at equivalent ΔK values the short cracks tend to grow faster than the long ones. This can be seen in Fig. 3 where the results from long cracks are included for comparison. Moreover, the short cracks can propagate below the threshold determined for long cracks. Consequently it follows that long crack data may provide an overly optimistic assessment of a materials fatigue crack propagation (FCP) resistance.

K_{op} is an increasing function of crack length and the value for long cracks is reached only when the small cracks extends over a distance of about 1,5 mm. The value for long cracks is 4.1 MPa√m. According to Pineau [6], K_{op} for small cracks is an increasing function of crack length. Pineau has shown that the data can be fitted to the function below:

$$K_{op} = 5.2a - 1.97a^2 + 0.24a^3 - 0.32$$

The derived formula is used in the computer code mentioned above. The program was designed to work in a way of cycle by cycle calculation, dN=1. One way of decreasing the computing time is to enlarge dN; this however introduces an error in the calculation. After some tests a decision was taken that all calculations should be made with *dN*=1000.

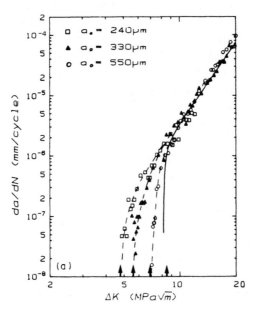

Figure 3. Crack growth rate as a function of *ΔK*. Plain curve for large specimens, dotted lines for small crack specimen [6].

Fatigue testing

For these tests a 4-point bending fixture was used. The distance between the outer jaws where 60 mm. Between the inner jaws (l=40 mm) the specimen is exposed to a constant bending moment. A large portion of the cast surface and the defects in connection with the surface is exposed to maximum load. All fatigue tests have been made with a load ratio of *R*=0. The stress amplitude has been held in the range 140-237 MPa. This is a normal range for chassis components where single peaks might reach 250 MPa.

RESULTS

Fatigue testing

The results from the fatigue tests are shown in Table 5. Specimens where the surface defect initiated fracture are marked with (*). On the other specimens fracture were caused by inclusions, slag stringers or micro shrinkage pores.

K. *Hamberg* et al.

Table 5. Amplitude stress, cycles to failure, size of initial defect, fatigue fracture and surface defect.

	σ_a	N_f	a_i	c_i	a_f	c_f	a_s	c_s
	MPa	10^3	mm	mm	mm	mm	mm	mm
D3A*	237	53	0,45	1,38	4,54	14,11	0,42	0,92
D1A	228	130	0,45	0,62	5,74	15,36	0,35	0,69
D6A	182	791	0,53	0,45	5,21	14,04	0,37	0,92
D17B	180	1.555	0,23	0,17	5,36	15,48	0,80	1,60
D15B	180	544	0,45	0,23	5,33	14,31	0,42	1,05
D14B	180	484	1,30	0,84	5,26	15,79	0,62	1,00
D10B	180	804	2,05	0,98	5,10	12,68	0,74	1,22
D13B	180	723	1,65	0,54	5,02	15,02	0,62	1,02
OK2A	180	1.206	0,75	0,31	4,55	10,31	-	-
OK6A	180	863	0,55	0,61	3,92	7,61	-	-
OK7A	180	956	0,21	0,20	4,31	8,37	-	-

Computer calculations

Table 6 shows the results from the computer calculations. The initial defect size, a_i, c_i, fatigue fracture size along with the stress amplitude are inputs to the program. The calculated fatigue life N_c is the output. Table 6 also shows the calculated effective initial stress intensity at the depth and the width of the defect.

Stress amplitude (σ_a), cycles to failure (N_f), calculated cycles to failure (N_c), calculation error (N_f/N_c), initial defect size (a_i and c_i) and initial stress intensity for computable specimens are shown in table 6. The stress intensity is computed in a and c directions (ΔKa_i and ΔK c_i.)

Table 6. Results from fatigue tests

	σ_a	N_f	N_c	N_f/N	a_i	c_i	ΔK_{ai}	ΔK_{ci}
	MPa	10^3	10^3	c	mm	mm	MPa$\sqrt{m}$	MPa$\sqrt{m}$
D3A	237	53	117	0,45	0,45	1,38	16,3	10,3
D1A	228	130	95	1,37	0,45	0,62	14,8	13,2
D6A	182	791	900	0,88	0,53	0,45	5,8	7,2
D17B	180	1.55	1.85	0,83	0,23	0,17	4,2	5,2
D15B	180	544	1.32	0,41	0,45	0,23	4,0	6,4
D14B	180	484	488	0,99	1,30	0,84	4,4	9,8
D10B	180	804	660	1,22	2,05	0,98	3,4	18,3
D13B	180	723	884	0,82	1,65	0,54	2,8	18,2
OK2A	180	1.20	1.54	0,78	0,75	0,31	4,4	11,9
OK6A	180	863	697	1,24	0,55	0,61	7,4	7,6
OK7A	180	956	1.76	0,54	0,21	0,20	4,8	5,1

As can be seen in Table 6 the factor that shows computed error is $0.41 < N_f/N_c < 1.37$, this can also be seen in Fig. 4 which shows the calculated lifetime, N_c versus cycles to failure, N_f for the computable specimens. The calculations are most often overestimated.

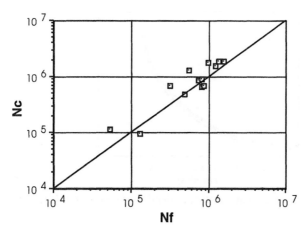

Figure 4. Calculated lifetime versus actual lifetime.

Computable specimens are specimens with a sizeable surface defect. An example of specimen that can not be used is a test bar with multiple starting points as the code stands today.

DISCUSSION

Fatigue Testing

The fatigue testing clearly showed that the surface defects initiated fracture only in one case, specimen D3$_A$. The most likely cause for this occurrence is the fact that the specimen contained no slag defects in the rim zone, only the surface defect. All cast components are shot blasted, which removes the dross or the sand grains and leaves a crater on the surface of the component. Underneath the crater compressive stresses are present and restrict crack initiation. Consequently, it is not necessary to reject a component only because of its visible surface defects, provided that the size of the defect is within normal foundry practice, i.e. $a < 0.8$ mm and $c < 2.5$ mm. Surface defects slightly larger than this will not reduce the fatigue life noticeably. This is a very important observation that can lower the number of rejections made from surface defects. The other type of defects in the rim zone, the slag or graphite stringer is more harmful. The depth of the stringer is so large that the compressive stresses can not penetrate at these depth, therefore the defect is most harmful. It is also important to develop more effective none destructive testing methods to sort out components containing crucial defects, such as different types of stringer. As can be seen from the results in table 6 the stringers grater than 0,5 mm in depth (a-value) is injurious.

With the designed computer program it is possible to investigate the behaviour of different parameters during the FCP calculation. Outputs from calculations made on specimen D10B be seen in Fig. 4, 5 and 6.

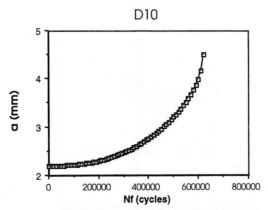

Figure 4. Calculated crack propagation in direction a (depth) for specimen D10B (a_i=2.03 mm, a_f=5.1 mm).

Figure 4 shows the FCP in direction a (depth). The gradient is low at the initiation stage of the FCP and increases as the crack propagates. Towards the end of the fatigue life the gradient increases rapidly and final fracture becomes apparent. The gradient of the FCP in direction c is larger at the initiation stage than after a number of cycles. This behaviour can also be studied in Fig. 5 where the stress intensity in direction c is large at first, $\Delta K_{c,eff}$ =18.3 MPa√m and then decreases to 13 MPa√m after which it begins to increase as the crack propagates. The reason for this is the configuration of the defect. From Table 6 it becomes apparent that the initiation defect in specimen D10B is more than twice as deep as it is wide. Observed behaviour can be expected because of multiple reasons. First a surface flaw tends to grow into a semicircle, at least in theory. Second the fact that for bending, the stress has its maximum at the surface. These two circumstances in combination with a short crack phenomena leads to the observed phenomena. Consequently the crack will propagate more in direction c, at the surface, than in direction a.

Figure 6 shows the calculated fatigue crack growth rate, da/dN as a function of the effective stress intensity. The typical appearance of Paris law prevails. The crack propagates below the threshold for long cracks, ΔK_{th} =7MPa√m, in this case the propagation rate is very small (10^{-7} mm/cycle equals 1Å/cycle). This propagation rate represents a mean FCP rate. At high stress intensities the function seems to deviate away from the expected behaviour. This is because the program has not at all been designed to take into account high stress intensities.

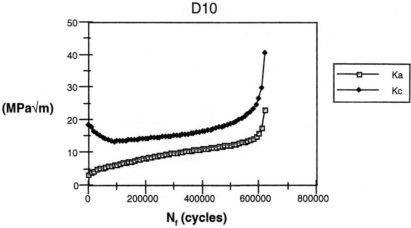

Figure 5. Variation of stress intensity factors in direction *a* and *c* through the fatigue life calculation for specimen D10$_B$.

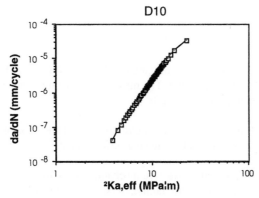

Figure 6. Calculated fatigue crack growth rate *da/dN* as a function of effective stress intensity factor range *ΔK* for specimen D10$_B$.

Results from the fatigue life calculations and experiments is shown in the form of a S-N-curve in fig. 7. Squared symbols marks the result of the actual fatigue test. The difference between calculated and actual test result is very small.

K. Hamberg et al.

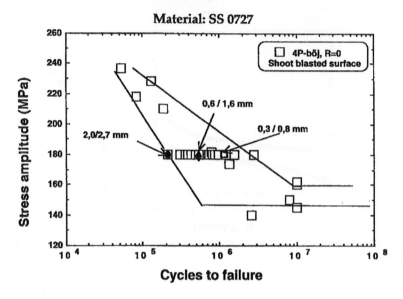

Figure 7. Calculations on some defects along with fatigue test results.

CONCLUSION

Visible surface defects in a magnitude of a = 0.8 mm and c = 2.3 mm, that often lead to rejection of components do not lower the fatigue life noticeable. The most likely reason for this is the compressive stress that is built in by the shot peening in these defects.

Dross defects and sand grains in connection with the as cast surface are the type of defects that most often lead to fracture, even if they are much smaller than the surface defects.

Fatigue crack propagation calculations based on fracture mechanics and Paris law in nodular cast iron can show good agreement with actual cycles to failure, if short cracks, stress ratio *R* and three dimensional constraint is taken into consideration.

It is possible to calculate the fatigue life for a specimen/component with the designed program within a reasonable time, even with normal computer power. This is done by a developed cycle by cycle calculation.

REFERENCES

1. Bradley, W.L., McKinney, K.E., and Gerhardt, P.C., Jr. (1986), *Fracture Toughness of Ductile Iron and Cast Steel, Fracture Mechanics: Seventeenth Volume. ASTM STP 905*, J.H. Underwood, R. Chait, C.W. Smith, D.P. Wilhem, W.A. Andrews, and J.C. Newman, Eds., American Society for Testing and Materials, Philadelphia,. pp 75-94.
2. Starkey, M. *The Influence of Microstructure on the Fatigue Properties of Spherodial Graphite Cast Irons*. (1980), GKN Group Technological Centre Memorandum No. 1492,.

3. Sofue, S. Okada, and T. Sasaki, (1978) *High Quality Ductile Cast Iron With Improved Fatigue Strength*, AFS Transactions, vol 86, , pp 173-182.

4. Newman, J.C., Jr and Raju, I,S,(1986), *Stress-intensity factor equations for cracks in three-dimensional finite bodies subjected to tension and bending loads*, Computational methods in the mechanics of fracture, *Vol. 2*, 311-333.

5. Elber, W,(1971), *ASTM STP486*, p.230.

6. Clement, P., Angeli, J. P., and Pinau, A. (1984). *Short crack behaviour in nodular cast iron*, Fatigue Engng Mater. Struct., Vol. 7, No. 4, pp. 251-265, 1984.

7. Owadano, T and Kishitake, K.(1984) *Fatigue Crack Propagation in Speroidal Graphite Cast Iron*, In: proceedings of 51.° International Foundry Congress,.

8. Fredheim, S.(1985) *Bruddmekaniske data for en del støpegods. Resultater av en litteraturundersøkelse*, Veritec,.

9. Suresh, s. and Ritchie, R. O. (1982) *Metal. trans.A*, 192, vol. 13A, , pp 1627.

10. Öberg, T and Wallin, K. (1986) *Fracture proporties of ferritic, ferritic-pearlitic, pearlitic and bainitic nodular cast iron*, Technical Researh cenyer of Finland, Research Reports 430.

11. Lankford, J. "The Growth of Small Fatigue Cracks in 7075-T6 aluminium", *Fatigue Engng Mater. Struct.* 5, 1982, 233-248.

12. Taylor, D. and Knott, J. F. "Growth of Fatigue Cracks From Casting Defects in Nickel-Aluminium Bronze", *Metals Technol.* 9, 1982, 221-228.

13. Schijve, J. "Differences between the growth of small and large fatigue cracks in relation to threshold K values", *Fatigue Threshold*, 1982, EMAS Publ., 881-908.

FATIGUE DESIGN AND QUALITY CONTROL OF FLAPPER VALVES

H. IKEDA

Toyoda Automatic Loom Works, Ltd. Kariya, 448 Japan

Y. MURAKAMI

Dept. Mech. Sci. & Engng., Kyushu University, Fukuoka, 812 - 81 Japan

ABSTRACT

Quality control of materials and final mass production parts from the viewpoint of development and fatigue design is performed using a combination of a new inclusion rating method, based on the statistics of extreme, and the $\sqrt{}$ area parameter model. The fatigue strength of mass production parts is predicted based on Vickers hardness and inclusion size. A reliable safety factor is defined by the proposed method.

KEYWORDS

Safety factor, fatigue limit prediction, statistics of extreme values, non-metallic inclusions, quality control, fatigue design

INTRODUCTION

Our company manufactures 6 million compressors for automobile airconditioners per year. Every customer requires higher reliability. Therefore, an improved method to guarantee the reliability of mass production parts is needed. Fatigue design and quality control methods that allow the development of new products with higher quality, lower cost and in a shorter time than present must be studied. The need to offer inexpensive products in a timely fashion is increasing.

For high strength steels used as flapper valves for the suction and discharge valves of compressors, non-metallic inclusions have a detrimental effect on fatigue strength and cause scatter in the fatigue strength due to inclusion size and position. Figure 1 is a photograph of a damaged valve. In Fig. 2 there are two flow charts, one is a conventional design flow chart and the other is a new fatigue design chart utilising the statistics of extremes and the $\sqrt{}$ area parameter model. The conventional fatigue limit, σ_{wexp}, was determined from experimental results on 10 to 20 pieces and the safety factor, S, was based on the stress during service loading, σ_{max}, using the relation S = ($\sigma_{wexp} / \sigma_{max}$). This method has two questions. One is that the scatter of fatigue strength is not considered. Another is that the safety factor is inaccurate and, consequently, parts tended to be over-designed with increasing cost. During mass production, the loss of balance between cost and quality is magnified.

In this paper, a new fatigue design method is proposed to predict the lower limits of the scatter of fatigue strength for 6 million samples. As the final result we get reliable and accurate safety factor for 6 million components.

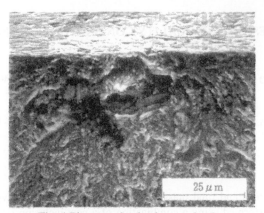

Fig. 1 Photograph of a damaged valve.

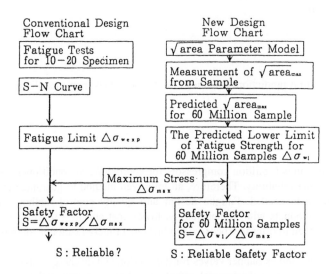

Fig. 2. Conventional and new fatigue design chart.

STRESSES ON THE SUCTION AND DISCHARGE VALVES

The suction and discharge valve operation models are given in Fig. 3. Pressure difference between the piston bore and the suction or discharge chamber opens or closes the valves. There are three types of stresses on the valves: (1) a bending stress on the root of the valves when the valves open, (2) a bending stress on the centre of the ports when the valves close, and (3) an impact stress during the impact moment. Cyclic loading in cases (1) and (2) is pulsating stress. In this paper we report the prediction method of the lower limits of the scatter of fatigue strength and safety factor for 6 million samples for cases (1) and (2) of a discharge valve.

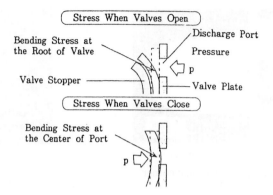

Fig. 3. Stress at discharge valve.

MATERIALS AND MEASUREMENTS

Materials

Chemical composition, surface roughness, residual stress, and mechanical properties are given in Tables 1 and 2. Figure 4 shows the conditions of specimen surface and corner, and Fig. 5 presents the microstructure.

Table 1. Chemical composition of the specimens (wt%)

sample	C	Si	Mn	P	S	Cr	Mo
A	1.00	0.19	0.44	0.006	0.006	0.13	≤0.01
B	1.00	0.23	0.44	0.008	0.010	0.21	0.03
C	1.00	0.23	0.45	0.007	0.007	0.13	≤0.01

Table 2. Surface roughness, residual stress, and mechanical properties

sample	roughness R_{max} (μm)	hardness Hv(5) (kgf/mm^2)	residual stress (N/mm^2)	σ_{ult} (N/mm^2)	elongation (%)	Young's modulus (N/mm^2)
A	1.00	562	584	1970	5.5	205000
B	0.92	566	639	1955	4.9	203000
C	0.90	562	621	1945	5.4	203000

Three metal samples were prepared by different factories. Then each sample was blanking-pressed from a strip and tumbled.

(a) side view of specimen   *(b) section of specimen*

Fig. 4. Conditions of specimen surface and corner.

Fig. 5. Microstructure.

Equipment and methods

Method of Bending Fatigue Test. Figure 6 shows the equipment used for the bending fatigue test. Cyclic loads are applied to the specimens by oil pressure, the magnitude of which is controlled by the oil pressure sensor. Figure 7 shows the relationship between oil pressure and stress. Maximum stress is determined by oil pressure based on Fig. 7. Figure 8 shows the shape and dimensions of the specimens. A sample size of 20 pieces was used and the fatigue tests were stopped at 5×10^6 cycles.

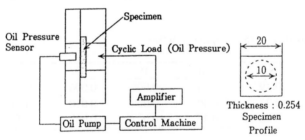

Fig. 6. Equipment for bending fatigue test.

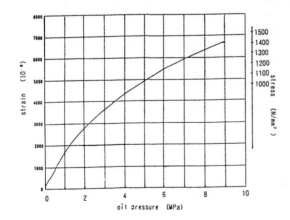

Fig. 7. Relationship between oil pressure and stress.

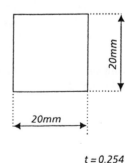

$t = 0.254$

Fig. 8. Shape and dimensions of specimen.

Measurement of √Area$_{max}$; Size of Inclusions [1 - 2]: Inclusion size is defined by the square root of the area projected on the plane perpendicular to the stress axis. The procedure for the measurement of √ area is as follows: a surface perpendicular to the direction of metal flow was polished with #2000 emery paper, and then finished with buff. The measurement of √ area$_{max}$ was done on 40 different places with the standard inspection area S$_o$, S$_o$ = 0.0064 mm^2. Three kinds of samples were checked, samples A - C. An additional 8 samples, A1 - A8, manufactured over a 2 year period at the same factory as sample A were also checked.

RESULT AND DISCUSSION

Result of Fatigue Tests and the Result of Measurement of √Area$_{max}$

Fatigue test results are shown in Fig. 9. Figure 9 indicates that the fatigue strength for all samples evaluated using a small number of test pieces are approximately the same. Figures 10 and 11 show the distributed of the maximum non-metallic inclusions in standard inspection area size, √ area$_{max}$, plotted on the sheet of statistics of extreme.

Comparison of the Experimental Fatigue Strength with the Prediction by the √Area Parameter

To predict the fatigue strength of mass production parts, the method based on statistics of extreme and the √ area parameter model is applied. The prediction method is as follows. First, the square root of the maximum projected area of inclusions, √ area$_{max}$ of these specimens is predicted from Fig. 10. Table 3 shows √ area$_{max}$ of inclusion for 43 samples of A, B and C.

Table 3. √ area$_{max}$ of inclusion for 43 samples of A-C.

sample	best fit line	N	V$_S$	T(N)	Y	√ area$_{max}$ (µm)
A	0.4828Y + 1.2145	43	0.316	83588	11.3336	6.69
B	0.5692Y + 1.2061	43	0.316	83588	11.3336	7.66
C	0.5749Y + 1.4107	43	0.316	83588	11.3336	7.93

The return period, T(N), for N samples is computed using

$$T(N) = N \cdot V_S / V_O \tag{1}$$

where V$_S$ is the critically stressed volume for the test specimens (range in $\sigma / \sigma_{max} \geq 0.9$, see Fig. 12), V$_O$ is the standard volume (0.0064 mm^2 × 0.1 × 0.254 mm = 1.6256 × 10^{-4}), and N is the number of specimens.

The return period T and standard reduced variate Y are related by

$$Y = -\ln (-\ln(T(N) - 1) / T(N)) \tag{2}$$

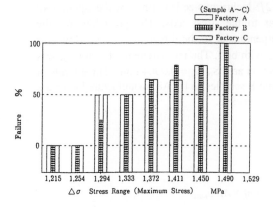

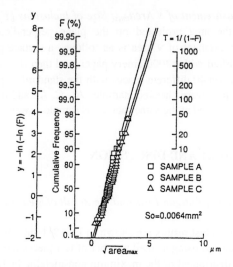

Fig.9. Result of bending fatigue tests

Fig.10. Distribution of the maximum nonmetallic
inclusions in standard inspection of
sample A-C

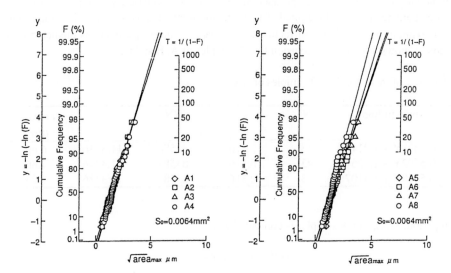

Fig.11. The distribution of the maximum nonmetallic inclusions
in standard inspection of sample A1-A8

The lower limit of the scatter of fatigue strength, σ_{wl}, is calculated by Eq. (3) [3, 4].

$$\sigma_{wl} = 1.41 \cdot (Hv + 120) / (\sqrt{area_{max}})^{1/6} \cdot ((1-R/2)^{\alpha} \tag{3}$$

where $R = \sigma_{min} / \sigma_{max}$, $\alpha = 0.226 + Hv \times 10^{-4}$, Hv is Vickers hardness, and $\sqrt{area_{max}}$ is for N samples and has dimensions μm.

Table 4 compares the results of fatigue tests and the predicted fatigue strength from Table 3 and Eq. (3) for each test piece. The predictions are in good agreement with the experimental results showing the validity of the proposed method.

Table 4. Comparison of the fatigue tests and the predicted fatigue strength

sample	$\sqrt{area_{max}}$ (μm)	Hv (kgf/mm^2)	residual stress (MPa)	$\Delta\sigma_{wl}$ (MPa)	$\Delta\sigma_{wexp}$ (MPa)	$\Delta\sigma_{wl}/\Delta\sigma_{wexp}$
A	6.69	562	-584	1350	1300	0.963
B	7.66	566	-639	1356	1300	0.959
C	7.93	562	-621	1336	1300	0.973

Lower Limits of the Scatter of Fatigue Strength and the Safety Factor for 6 Million Parts

To predict the lower limits of the scatter of fatigue strength, we consider the lower limits of the hardness and residual stress. The equation for the best fit line for C is used because sample C has the largest $\sqrt{area}$ among samples A-C. Table 5 shows the predicted $\sqrt{area_{max}}$. Table 6 shows the lower limits of the scatter of fatigue strength and the safety factors defined by ($\Delta\sigma_{wl}/ \Delta\sigma_{max}$) in conventional and new design method. The number of parts is assumed N = 60 million, because each compressor has ten valves. It has been empirically recommended that the safety factor of the functional parts to be 2.0 in the conventional design, because we can not predict all scatter. The safety factor on the position of valve centre in Table 6 is 1.66 in the conventional design. Based on the conventional design concept for this case, we had judged it was necessary to have a higher fatigue strength. With the new fatigue design method, the safety factor is 1.41. That is lower than the conventional one, but this is an accurate and reliable safety factor. We can conclude that the safety factor of the new fatigue design method is useful because we have had no trouble with this valve over the past ten years. Applicability of the new fatigue design method to future productions is supported.

Table 5. Predicted $\sqrt{area_{max}}$ for 60 million parts.

situation	best fit line	N	V_V^*	T(N)	Y	$\sqrt{area_{max}}$ (μm)
(1) root of the valve	0.4828Y + 1.2145	6×10^7	1.8300	5.625×10^{11}	27.0556	16.96
(2) centre of the valve	0.5749Y + 1.4107	6×10^7	0.2422	7.455×10^{10}	25.0334	14.39

* V_V is the critical volume for the valve defined in Fig. 12.

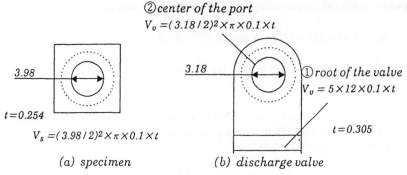

Fig. 12. Critical volume

Table 6. Lower limit of fatigue strength and safety factor for 60 million parts.

design method	situation	$\sqrt{area}_{max}$ (μm)	Hv	residual stress (MPa)	$\Delta\sigma_{wl}$ (MPa)	$\Delta\sigma_{wexp}$ (MPa)	$\Delta\sigma_{wl}/\Delta\sigma_{wexp}$
new design	root of the valve	16.96	545	-399	1084	490	2.21
	centre of the valve	14.39	545	-399	1110	785	1.41
conventional design	root of the valve	-	-	-	1300	490	2.65
	centre of the valve	-	-	-	1300	785	1.66

Scatter of Production

To investigate the scatter of production for 500 thousand units for one month production, $\sqrt{area}_{max}$ of sample A1 - A8 is predicted from Fig. 11, and results are shown in Table 7. We have confirmed that the scatter of the quality of mass production is small, because $\sqrt{area}_{max} = 16.54$ μm is less than 16.96 μm that is the predicted upper limits of the scatter of inclusion size for 6 million units.

Table 7. Scatter of mass production ($V_V = 1.83$, $T(5\times10^6) = 4.687\times10^{10}$, $Y = 24.5706$).

sample No.	product month	best fit line	$\sqrt{area}_{max}$ (μm) root of the valve
A1	92/5	0.6034Y + 1.2354	16.06
A2	92/10	0.5424Y + 1.3545	14.68
A3	93/2	0.5901Y + 1.2110	14.50
A4	93/12	0.5876Y + 1.3248	15.76
A5	94/3	0.5336Y + 1.2670	14.38
A6	94/5	0.5474Y + 1.4127	14.86
A7	94/10	0.6210Y + 1.2830	16.54
A8	94/11	0.4828Y + 1.2145	13.08

CONCLUSION

The $\sqrt{}$ area parameter model and the inclusion rating method based on statistics of extreme were applied to the fatigue design of mass production parts; compressor flapper valves. The conclusions are summarised in the followings;
1. The predicted fatigue strength by the $\sqrt{}$ area parameter model is in good agreement with the experimental result.
2. An accurate and reliable safety factor was determined by the new fatigue design method considering the scatter of fatigue strength caused by the scatter of inclusion size and position. The new fatigue design method is very useful to confirm the fatigue reliability of mass production parts.
3. The inclusion rating based on the statistics of extreme is useful for the quality control of materials used for mass production parts.

REFERENCES

1. Murakami, Y. "Inclusion Rating by Statistics of Extreme Values and Its Application to Fatigue Strength Prediction and Quality Control of Materials". Journal of Research of the National Institute of Standards and Technology, Vol. 99, No. 4, July - August 1994, pp. 345 - 351.
2. Murakami, Y., Toriyama, T., and Coudert, E. M. "Instructions for a New Method of Inclusion Rating and Correlations with the Fatigue Limit". Journal of Testing and Evaluation, JTEVA, Vol. 22, No. 4, July 1994, pp. 318 - 326.
3. Murakami, Y. "Metal Fatigue: Effects of Small Defects and Non-metallic Inclusions". Yokendo, Tokyo, 1993.
4. Murakami, Y., Endo, M. "Effect of Defects Inclusions and Inhomogeneities on Fatigue Strength". Int. J. Fatigue, 1994, Vol. 16, April, pp. 163 - 182.

DURABILITY ASSESSMENTS OF STRADDLE CARRIER WHEEL SUPPORTS

A. SILJANDER, A. KÄHÖNEN, M. LEHTONEN
VTT Manufacturing Technology, P. O. Box 1705, FIN 02044 VTT, Finland

P. MÄNTYSAARI
Sisu Terminal Systems, Inc., P. O. Box 387, FIN-33101 Tampere, Finland

ABSTRACT

This paper describes the analytical and experimental durability assessments of a straddle carrier wheel supports. Field measurements were conducted and the multiaxial road inputs were quantified for subsequent Finite Element analyses and laboratory service simulation tests to compare two different design alternatives. The results of this study will be implemented in the next generation of wheel supports.

INTRODUCTION

In 1993, VTT Manufacturing Technology and representatives from Finnish and Swedish transportation and steel industries initiated a research project KONSPRO *("Optimised Constructions Using High Strength Weldable Steels")*, which is a part of an inter-Nordic research program NORDLIST (Nordic Research and Development of Light Structures). The Finnish part of the ongoing KONSPRO-project *("Optimisation of Weight and Endurance of Vehicle Structures")* is focused on on-road and off-road transportation industry. Among the primary objectives of the Finnish project are: *i)* to reduce vehicular weight via the use of higher strength weldable materials while ensuring adequate fatigue strength; *ii)* to expand the conventional hot spot-method [1-3] to multiaxial loading situations and *iii)* to integrate the new multiaxial hot spot-approach to an existing FE-post-processor. The main benefits to be gained from the project results are: *i)* enhanced structural reliability, *ii)* reduced material and production costs and *iii)* improved technological competitiveness.

Extensive and systematic field measurements have been used to capture the "duty cycles" [e.g. 4] and to quantify multiaxiality of the structural nominal and hot spot strain responses. These field measurements form a realistic basis for subsequent and on-going analyses, which are carried out for the case-studies of the participating Finnish industry: *Sisu Defence Oy* (heavy-duty off-road truck axle; welded construction), *Sisu Terminal Systems Inc.* (straddle carrier; cast wheel support) and *Oy Närko Ab* (freight trailer frame & bogie; welded constructions). Selected case-study components are also being fatigue tested using multiaxial and phase-related recorded road-load data [5].

Additional fatigue tests will be conducted on a rectangular hollow section tube-to-plate welded cantilever beam-type geometry manufactured by *Rautaruukki Oy Metform*. The baseline fatigue test data will be gathered by conducting a series of bending only and torsion only constant amplitude fatigue tests. Once the degree of multiaxiality during typical duty-cycle events has been experimentally documented from the above case-studies, the in-phase and out-of-phase combined bending-torsion constant amplitude fatigue tests will be conducted using the cantilever square hollow section tube-to-plate as-welded component [6]. This paper briefly describes one of the case studies within the Finnish project concerning the cast wheel support of a straddle carrier manufactured by *Sisu Terminal Systems Inc.* of Tampere, Finland. The study was aimed at quantifying the road inputs for ˙subsequent structural analyses and laboratory service simulation tests to analytically and experimentally compare two different design alternatives.

OVERVIEW OF THE WHEEL SUPPORT

Following e.g. the unloading of the cargo containers from overseas cargo vessels using either quayside cranes (*Lo-Lo* ships) or terminal tractors (*Ro-Ro* ships), the containers are temporarily stacked within the perimeter of container yards of harbour cargo terminals for subsequent road/rail freight transportation. Figure 1 shows an overview of a straddle carrier. Vehicles of this type are commonly used in harbour environments to pick, carry and stack cargo containers.

Fig. 1. An 8-wheel steer, 4-wheel drive straddle carrier in operation by a quayside crane in the Port of Helsinki.

The engines within the side beams of the vehicle supply the torque to the inner two wheels on both sides of the vehicle. Each steerable torque-powered wheel is connected to an adjacent free running and also steerable wheel. The cast wheel support is one of the primary load carrying components of the straddle carrier. Because its fatigue strength is of primary concern for the reasons of safety and uninterrupted production, the wheel support was selected as the object of this study. The wheel support sustains a variety of operational loads (Fig. 2): A vertical wheel load (F_{VER}), resulting from the container lifting and rough surface (railroad tracks, pallets, potholes etc.) manoeuvring operations; a longitudinal wheel load (F_{LON}), resulting from acceleration and braking operations; and a lateral wheel load (F_{LAT}), resulting from left and right turns of the vehicle. The support reaction forces within the wheel support (from the bearing support points of the chain-type and gear wheel-type final drives) were ignored in the analyses.

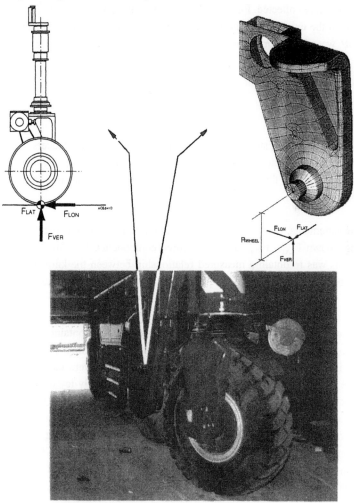

Fig. 2. Schematic of the cast wheel support and the assumed external operational wheel load components.

AIM AND SCOPE

The ultimate goal of this case study was to experimentally verify the adequate fatigue strength of a new design concept of the wheel support (i.e. gear wheel-type final drive) without increasing component weight or manufacturing costs and while maintaining interchangeability with existing models of the wheel support (i.e. chain-type final drive). To ensure the adequate fatigue strength of the new wheel support, an integrated design strategy [7] involving both experiments and analysis was implemented.

Within the framework of the above requirements, field measurements were conducted with the existing cast component and operational duty cycle data combined with static component calibrations for subsequent analyses were collected. Finite element (FE) analyses were then performed to quantify the differences between the existing and the new design of the wheel support. Finally, both designs of the wheel support were laboratory proof tested to experimentally verify the analytical analysis results.

FIELD MEASUREMENTS

Strain gages *(s)*, accelerometers *(a)*, steering angle angular displacement transducers *(<)* and pressure transducers *(p)* were fixed to selected locations of the vehicle. All transducer signals were recorded with an instrument recorder such that all instrumentation hardware was installed in the operator cab of the straddle carrier, Fig. 3a.

Prior to the field measurements, extensive component calibrations were implemented by loading the wheel support using known F_{LAT}-, F_{VER}- and F_{LON}- force components, Fig. 3b. The purpose of the component calibrations was to obtain a measured relationship between the known input forces and the resulting responses (ε_{VER}, ε_{LAT}, ε_{LON}) for parallel FE-analyses and subsequent laboratory proof tests.

Field measurements were collected during typical service usage on the container yard of the Port of Helsinki. Different operators were employed during the field measurements to include operator-induced differences in the measured responses [8]. The duty cycles were captured during winter conditions to include the "stick-slip" -type tire inputs due to occasionally icy driving surfaces (Fig. 1). Each duty cycle consisted of the following steps: **i)** manoeuvring without a container into close proximity of a new container by the quayside crane; **ii)** lifting a container to 1st floor using the carrier's own hoist system and the telescopic spreader; **iii)** manoeuvring container onboard (including brakings, accelerations, curves, railroad crossings, snow banks, pallets etc.) to a free location within the harbour's container yard; **iv)** stop manoeuvring while container onboard; set the container to 1st, 2nd or 3rd floor (i.e. on top of other containers); and **v)** repeat step (i).

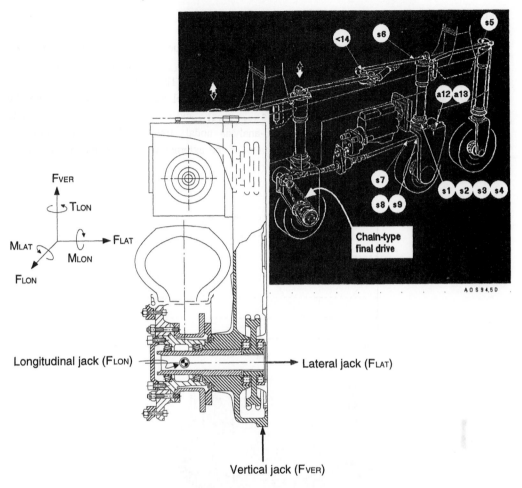

Fig. 3. (a) Overview of the field measurement set-up showing some of the instrumentation; (b) schematic of the static component calibrations prior to duty cycle documentation in actual field conditions.

During the field measurements that consisted of approximately 100 containers, the individual container weight data (when available) was real-time radioed from the command centre of the Port of Helsinki to the operator cab and subsequently read to the voice channel of the recorder together with relevant manoeuvring events. The average duration of a duty cycle containing of the above steps (ii)-(iv) was approximately 4 minutes.

In addition to the captured response signals $(\varepsilon, a, <, p)$, statistics of the container weight distribution, number of left/right turns, brakings, accelerations and full stops were among the data produced. The data were then compared to the database of Sisu Terminal Systems Inc. such that a representative duty cycle for the cast components' laboratory proof tests could be constructed.

FINITE ELEMENT (FE) ANALYSES

All FE-models were created using PATRAN pre-processor [9]. Due to the experimentally observed multiaxial and out-of-phase nature of the operational loads (F_{VER}, F_{LAT}, F_{LON}), three-dimensional solid element models were created from the instrumented components. The first model was created from the steering arm (connecting the wheel support to the carrier steel frame) to analytically study the nodal strain distribution in the vicinity of the ε_{VER}, ε_{LAT}, and ε_{LON} -strain gages of the field measurements, Fig. 4. By comparing the FE-analysis nodal strain results to the component calibration measured strain results, initial estimates of the external wheel load magnitudes (F_{VER}, F_{LAT}, F_{LON}) were obtained.

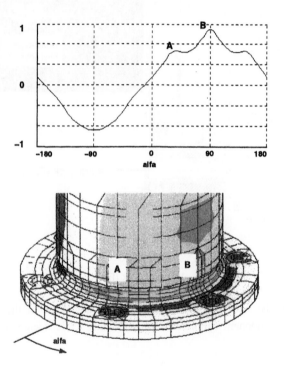

Fig. 4. An example of the FE-model and the analysis results of the steering arm.

Also three-dimensional FE-models of the entire wheel support were created using shell elements (coarse model of the chain-type final drive design), and a 20-node quadratic brick element model of the gear wheel-type design. The purpose of the FE-models was to identify most critical regions of the wheel supports, i.e. maximum stress regions. Again, nodal strain quantities were compared to the measured quantities captured during the component calibrations and duty cycle measurements. In these FE-models, there were approximately 8 000 nodes, 1 400 elements and 24 000 degrees of freedom. ABAQUS/Standard structural analysis code [10] was used for the analyses. An example of the FE-analysis results from one combined load case is presented in Fig. 5.

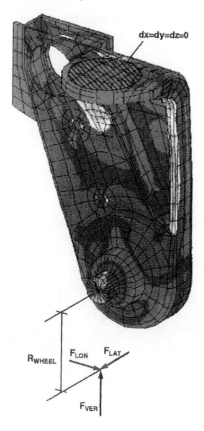

Fig. 5. Overview of an FE-model of the gear wheel-type final drive cast wheel support, boundary conditions and a schematic of the assumed wheel loads and the analysis results for one combined loading case.

SERVICE SIMULATION FATIGUE TESTS

A service simulation test system was built onto the tension floor of VTT Structural Engineering Laboratory. There were three orthogonally placed hydraulic actuators such that each actuator produced one individual wheel load component (F_{VER}, F_{LAT}, F_{LON}), Fig. 6. Due to the actuator swivel head mounting arrangement on the wheel spindle, all longitudinal wheel loads (F_{LON}) were transmitted as "±braking forces". For this reason the bearing support reactions within the cast component were ignored in the comparison FE-analyses. To simulate realistic boundary conditions, the wheel support was bolted to the steering arm flange (Fig. 4), which was attached inside a rigidly mounted steel box, which in turn simulated the stiffness of the steel frame of the straddle carrier.

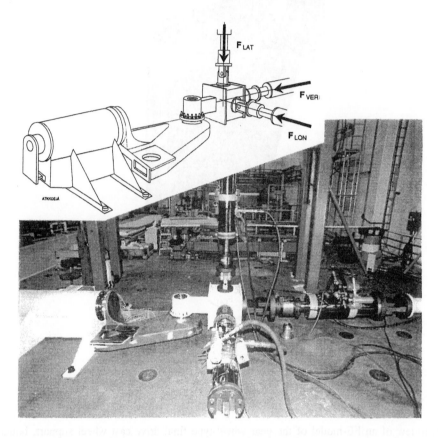

Fig. 6. Overview of the laboratory proof test set-up for the fatigue tests of the wheel supports. A similar two-actuator service simulation test system was employed in the experimental durability assessment concerning the heavy-duty off-road axle housings of Sisu Defence Oy.

Prior to the final proof tests, a series of component calibrations were again performed to compare the results to those conducted experimentally in field conditions (Fig. 3b) and analytically (FEA) in the laboratory (Figs. 4 and 5). The laboratory duty-cycle was based on selected typical events from the field measurements. The final laboratory duty-cycle contained 336 hauling operations such that a representative number of left/right turns, brakings and accelerations with and without containers of a desired statistical weight distribution were included.

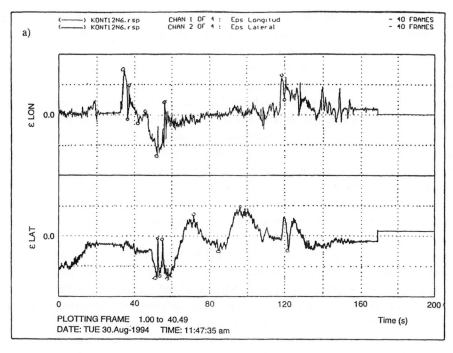

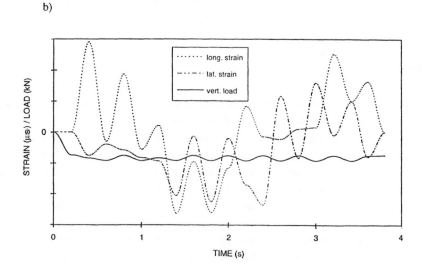

Fig. 7. (a) An example of the original response strain time histories captured during the field measurements (ε_{LAT} & ε_{LON}); (b) the corresponding compressed (peak sliced) and desired strain time histories for the CRPC iteration scheme.

During the field measurements, the average container handling duration was approximately 4 minutes. In order to carry out the laboratory proof tests within a reasonable testing time, the average container handling duration was reduced to approximately 4 seconds per hauling operation. The laboratory duty-cycles were compressed by removing all smooth driving events from the measured strain responses. In other words, only the original large-amplitude strain responses from typical and often experienced driving events were considered (Fig. 7a), which were then compressed using a peak slicing technique similar to that reported elsewhere [11]. In the manual peak slicing technique, the absolute maxima and minima per strain channel were stored while simultaneously storing the instantaneous strain values from the other strain channels of interest. A half sine wave was then fitted between the selected strain-time points. As a result, the desired strain responses for the CRPC (Component Remote Parameter Control) iteration scheme [12] were obtained, Fig. 7b. Significant time savings were obtained; over 10 years' service usage of the wheel support could be simulated in approximately 2 months of continuous testing time (i.e. 24 hours per day).

To iterate the hydraulic actuator-induced external wheel load time-histories in the laboratory, strain gages were instrumented to identical locations of the cast wheel support in view of the field measurement strain gage locations. The iterated wheel load and the resulting strain time-histories were then Rainflow counted [13], compared to those of the original field measurement results and used in fatigue life predictions. Magnetic particle method was among the methods employed to monitor possible surface crack formation throughout the load controlled proof tests. Both proof tests were continued beyond the design fatigue life to check the reserve limit e.g. for the purposes of maintenance period determination.

DISCUSSION

During the contract operation of the instrumented straddle carrier, there was no time to load road/rail freight cars for subsequent transportation. Instead, all containers were briskly hauled away from the quayside crane and piled to a temporary location. In view of the experimentally captured operational responses of the wheel support, manoeuvring without a container produced comparable responses to those conducted while container onboard (there are as many manoeuvring events with and without a container onboard)! A container onboard prevents the operators from racing in a similar manner as commonly occurs while manoeuvring without a container (this was manifested by a frequent warning tone of the standard safety unit monitoring the vehicle's accelerations). Further, the operators followed the manufacturer's instructions in that the containers were always held at the 1st floor during manoeuvring events. For these reasons, the laboratory duty-cycle consisted of events with and without a container onboard on the 1st floor. The laboratory duty-cycle therefore represented the normal service usage of the straddle carrier.

In contrast to the laboratory duty-cycle, there are many harbours where containers are held at the 2nd or even 3rd floor during manoeuvring operations, which immediately affects to the magnitude and distribution of the measured responses and therefore to the experimental and analytical fatigue life, not to mention the safety aspects. Therefore, additional field measurements (preferably longer in duration) should be conducted in these extreme conditions to capture the "1% driver" [8]. The known container weights should also be emphasised, since only approximately 50% of the containers carried during the field measurements were of known weight.

The frequency content (PSD) of the strain responses during the final laboratory duty-cycle (test spectrum) was similar to that captured in field conditions (service spectrum). Further, owing to the nature of the CRPC scheme, the maximum ranges of the Rainflow-counted strain responses were similar between the test and service spectra. However, due to the peak slicing technique employed, the number of large strain cycles in the test spectrum was somewhat higher than that of the service spectrum. For the same reason, the test spectrum contained a smaller number of small and intermediate strain ranges than the service spectrum. Therefore, the shape of the test spectrum was more severe than the service spectrum, but the maximum strains were similar between the two spectra. The fatigue life predictions of the wheel support using the test spectrum and the service spectrum are underway. Also, the cast material's small-scale specimen fatigue test data are being analysed using fracture mechanics.

CONCLUSIONS

This paper has outlined the integrated design strategy [7] to experimentally and analytically compare two design alternatives of a cast wheel support of a straddle carrier. Conclusions of this study can be summarised.

(1)   Field measurement results from the normal service usage of an existing straddle carrier wheel support (chain-type final drive) provided the basis for realistic durability assessments.

(2)   Structural analyses using FE-analysis and service simulation fatigue tests were essential in comparing both analytically and experimentally the durability of chain-type and gear wheel-type final drive wheel supports subjected to identical multi-channel and phase-related road data.

(3)   Both designs of the wheel support exceeded the analytical and experimental durability requirements defined by Sisu Terminal Systems Inc.

(4)   The results of this case-study will be implemented in future design such that the existing chain-type final drive will be replaced by the gear wheel-type final drive in the next generation of cast wheel support components.

ACKNOWLEDGEMENTS

The durability assessments of the straddle carrier wheel supports were a part of an on-going KONSPRO-research project, which is a part of an inter-Nordic research program NORDLIST. Funding for the project is provided by *NI* (Nordisk Industrifond; Nordic Industrial Foundation) *TEKES* (Tekniikan edistämissäätiö; Technology Development Centre of Finland*), Sisu Defence Oy,*

Oy Närko Ab, Rautaruukki Oy Metform and *Sisu Terminal Systems Inc.* All sponsors are gratefully thanked for their financial support.

Mr. P. Hasanen (Finnsteve Oy Ltd., Port of Helsinki) is thanked for arranging the availability of the straddle carrier for the field measurements. Mr. P. Tuononen and Mr. T. Teittinen (VTT Manufacturing Technology) expertly carried out the transducer instrumentation and subsequent field measurements. Special thanks are extended to Mr. J. Juntunen (VTT Structural Engineering) for designing the experimental set-up for the laboratory fatigue tests and for conducting the proof tests. Mr. Gary Marquis (VTT Manufacturing Technology) is thanked for writing the computer program to produce the peak sliced data. Mr. M. Bäckström (VTT Manufacturing Technology) is acknowledged for conducting the Rainflow analyses during the service simulation tests. Mr. I. Pitkänen, Mr. M. Hirn and Mr. M. Kankare (VTT Manufacturing Technology) conducted the magnetic particle inspections throughout the fatigue tests. Ms. H. Hänninen and Mr. T. Hokkanen (VTT Manufacturing Technology) prepared the graphics for this paper.

REFERENCES

1. Niemi, E. (1993) Aspects of Good Design Practice for Fatigue Loaded Welded Components. In: *Fatigue Design, ESIS 16* (J. Solin, G. Marquis, A. Siljander, and S. Sipilä, Eds), Mechanical Engineering Publications, London, pp. 333-351.

2. Niemi, E. (1992). Recommendations Concerning Stress Determination for Fatigue Analysis of Welded Components. *IIW Doc. XIII-1458-92/XV-797-92.* International Institute of Welding, Abington Hall, UK.

3. Marquis, G., and Kähönen, A. (1995) Fatigue Testing and Analysis Using the Hot Spot Method. Espoo: *VTT Publications No. 239.* Technical Research Centre of Finland (37 p).

4. Martz, J.W., Smiley, R.G., and Kormos, J.G. (1978). Field Testing of "Reference Vehicles" as an Aid to the Design Analysis Process for Earthmoving Equipment. *SAE Paper No. 780485.* Society of Automotive Engineers, Inc., Warrendale, PA, USA.

5. Wright, D.H. (1993). *Testing Automotive Materials and Components.* Society of Automotive Engineers, Inc., Warrendale, PA, USA (243 p).

6. Siljander, A. (1995). Multiaxial Fatigue Damage Parameters - Literature Overview. Traditional Stress-Based Models versus Stress-Based Critical Plane Approaches. Espoo: *VTT Research Notes No. 1603.* VTT Manufacturing Technology (29 p).

7. Siljander, A., Lehtonen, M., Marquis, G., Solin, J., Vuorio, J., and Tuononen, P. (1993). Fatigue Assessment of a Cast Component for a Timber Crane. In: *Fatigue Design, ESIS 16* (J. Solin, G. Marquis, A. Siljander, and S. Sipilä, Eds). Mechanical Engineering Publications, London, pp. 321-331.

8. Shütz, W. (1993). The Significance of Service Load Data for Fatigue Life Analysis. In: *Fatigue Design ESIS 16* (J. Solin, G. Marquis, A. Siljander, and S. Sipilä, Eds). Mechanical Engineering Publications, London, pp. 1-17.

9. PATRAN. *User's Manual, version 2.5-4.* PDA Engineering.

10. ABAQUS Standard. *User's Manual, version 5.4.* Hibbitt, Karlsson & Sorensen, Inc.

11. Schmidt, R. (1993). Random fatigue testing based upon RPC III tools. In: *13th RPC User Group Meeting,* 26-27 May. Renault Lardy Test Center, Paris, France.

12. Component Remote Parameter Control. *User's Manual, version 3.20.* MTS Systems Corporation, Minneapolis, USA.

13. Kähönen, A. (1994). Comparison of Rainflow Classification Results Between VTTVFLOW and SOMAT-programs. Espoo: VTT Manufacturing Technology, *Report No VTT VALB-35* (27 p).

FATIGUE DESIGN OF PRESSED STEEL SHEET

ANDERS GUSTAVSSON and ARNE MELANDER
Swedish Institute for Metals Research
Drottning Kristinas väg 48
S-114 28 Stockholm
Sweden

Abstract

The present work is devoted to the fatigue design of pressed steel sheet structures as used in e.g. auto bodies. In pressed parts areas exist with more or less sharp curvatures. Characteristics of the fatigue process for such parts are crack inititiation in a highly strained area due to the forming of the sheet and a loading mode that is predominantly bending of the thin sheet.

The topic is addressed from two viewpoints. In the first instance, fatigue test procedures are discussed and results are presented for a common deep drawing steel and a high strength dual-phase sheet steel. Secondly, fatigue life estimation based on finite element simulation of both the forming and the following fatigue loading are presented.

The results indicate an advantage of using higher strength steels for pressed parts. Furthermore, the local strain aproach is reliable for fatigue life estimations. However, the predictions are improved by considering also fatigue crack growth.

Keywords: fatigue, sheet steels, finite elements, sheet metals forming, local strain approach

1. INTRODUCTION

With the overall aim to reduce panel weight of sheet auto body components, shallow pressed parts, often referred to as coinings, are introduced in order to either increase the panel stiffness or to facilitate attachments such as welds or nuts. Fig. 1 shows an example of such a pressed panel. However, the introduction of such pressings also raises the question of fatigue. This is so since the pressing operation as such influences the fatigue properties of the gauge material by thickness reduction and strain hardening. Furthermore, the formed geometry often introduces fatigue critical stress concentrations in the panel where a transition takes place from a pressed coining to the plane sheet.

73

A. Gustavsson and A. Melander

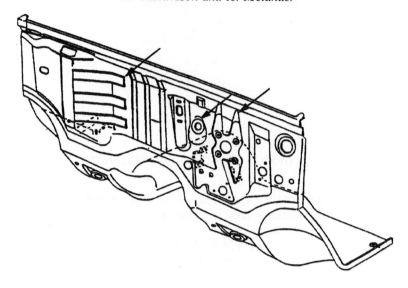

Fig. 1. *Example of a pressed panel. The arrows indicate several coinings as discussed in this paper.*

By the forming operation the sheet is locally strained as much as up to about 20% in the end portion of the coining where fatigue cracking is liable to occur. The effect on the fatigue properties of such levels of prestraining has been investigated earlier [4,7,8,13,17]. The conclusion from these investigations was that the gain in fatigue strength was proportional to that of the tensile strength if high degrees of prestraining was considered. Increases of the tensile stength by some 20-30% have been noted. However, for small degrees of prestraining, i.e. less than about 10%, small increments in fatigue strength were reported although considerable gain in tensile strength can be noticed [7].

This paper discusses the influence of geometrical stress concentration and forming strains of pressed sheet panels on the fatigue strength. . In the first instance the test method is introduced and fatigue life results are presented for a conventional deep-drawing sheet steel (DDQ) and a high strength dual-phase sheet steel (DP400). Next, finite element simulations of both the forming operation and the subsequent fatigue loading are discussed. Finally, with the finite element simulations as a basis fatigue life predictions for the two sheet steels tested are presented. Both predictions for the fatigue crack initiation and the fatige crack propagation stages are discussed.

2. MATERIALS AND EXPERIMENTAL INVESTIGATION

The two materials investigated were a conventional deep drawing sheet steel (DDQ) and a high strength dual-phase sheet steel (DP400). The sheet thickness was about 1.2 mm. The chemical composition is given in Table 1. The tensile properties were for DDQ $R_{P0.2}$=188 MPa, R_m=314 MPa and A_{15}=0.47. For DP400 $R_{P0.2}$=267 MPa, R_m=444 MPa and A_{15}=0.43.

Table 1. Chemical composition of investigated materials (in weight per cent)

Material	C	Si	Mn	P	S	N	Cr	Ni	Cu	Al
DP400	0.069	0.01	0.19	0.08	0.012	0.003	0.03	0.010	0.03	0.044
DDQ	0.053	0.004	0.212	0.011	0.018	0.004	0.015	0.021	0.008	0.049

Fig. 2. Photograph of the upper side the fatigue test specimen. The length of the specimen was 300 mm and the width 91 mm. At the arrows at the end portions of the coining fatigue cracks appeared as desired.

The fatigue specimen used in the present investigation is shown in Fig. 2. The specimens were formed using a conventional servohydraulic sheet forming press with the tool geometry displayed in Fig. 3. This tool has a punch and two dies which allows a blank holder pressure to be applied. During forming the specimen is pressed to a depth of about 5.7 mm. The surfaces of the sheet were lubricated using conventional deep-drawing oil. After forming the specimens were heat treated at 190°C for 15 minutes which simulates the paint baking operation of the auto panels.

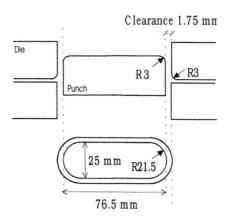

Fig. 3. Dimensions of the tools used to press the fatigue specimens.

The fatigue tests were performed in fully reversed four point bending in the test set-up shown in Fig. 4 [5,6]. The specimen were clamped at four locations where the distance between the two inner supports was 150 mm. Between the inner and outer support was a gap of 40 mm.

Fig. 4.   Fatigue test set-up for fully reversed four point bend testing of coined sheet samples.

In order to monitor the bending moment applied to the specimen the bending strains were recorded by applying a 5 mm long strain gauge half-way between the inner support and the end portion of the coining. This measurement provides a non-linear relation between the applied load and the nominal bending strain ε_{nom} due to pick up of bending moment in the supporting points [5]. The bending moment of the specimen can be calculated from conventional bending theory as

$$M_{nom} = \frac{b \cdot t^2 \cdot E \cdot \varepsilon_{nom}}{6 \cdot \left(1 - v^2\right)} \tag{1}$$

where b is the specimen width, t the sheet thickness and E=210 GPa is the Youngs' modulus.

For the determination of fatigue failure a certain stiffness criterion was used. This criterion was based upon the continous recording of the applied load (constant) and the simultaneous stroke of the loading piston. When the fatigue crack appears the stiffness of the specimen is reduced. The criterion was so formulated that when the stiffness had decreased by 3% from the stable level as recorded up to about half the fatigue life, then fatigue failure was defined to occur. This check was performed after the test was conducted in order to avoid false stops of the tests due to occasional peaks in the recorded signals. Fatigue cracking was found to initiate at the end portion of the coining. The crack initiated on the upper side of the sheet, Fig. 2, and gradually grew through the thickness and along the coining perifery. By inspection of the crack length at a few specimens it could be established to be about 20 mm at the fullfilment of the failure crieria.

Fig. 5 shows the fatigue life curves for the two materials investigated in terms of nominal bending moment vs number of cycles to failure (3% stiffness drop criterion). Clearly, the benefits of the higher tensile strength of DP400 also are maintained throughout the fatigue testing range considered here (10^4 to 10^6 cycles). The increment in fatigue life is about a factor of 4 for DP400 compared to DDQ.

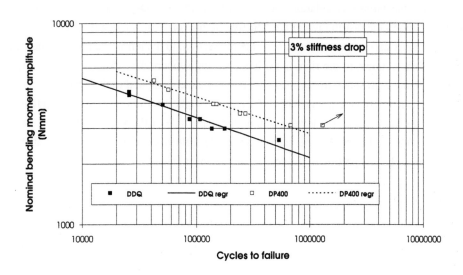

Fig. 5.   Four point bend fatigue test results for DP400 and DDQ.

3. FINITE ELEMENT SIMULATIONS

For simulation of the forming operation an elastoplastic material description based on a non-linear kinematic hardening model was used [5,10]. The material parameters used were obtained by numerical fitting to tensile test data. For the fatigue loading case only elastic calculations are presented here.

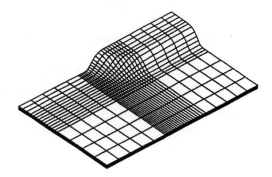

Fig. 6.   Finite element mesh as seen after forming simulation.

A finite element mesh consisting of one layer of 24 nodes brick elements were used. The quarter model used due to symmetries is shown in Fig. 6 after the simulation of the forming stage. In the subsequent fatigue loading a uniform bending moment was applied to the edge of the mesh corresponding to the inner support of the test set-up.

The results for the fatigue calculations are only presented for DDQ. The results for DP400 were similar. Fig. 7 shows the distribution of the effective strain (von Mises) along the x coordinate running along the symmetry axis of the specimen from the midsection; to the left in the figure is the midsection and the top of the coining, to the right is the inner loading support at x=75 mm. The two strain peaks shown in the figure correspond to the bend on the top of the coining and to the bend at the end portion of the coining where fatigue cracks occured. Clearly, the highest forming strains were obtained on the upper side of the top bend. In the lower bend, on the upper side of the sheet the forming strains were less than 0.05. This indicates that we cannot expect big influences of strain hardening at the fatigue critical location at the end portion of the coining on the upper side of the sheet [7].

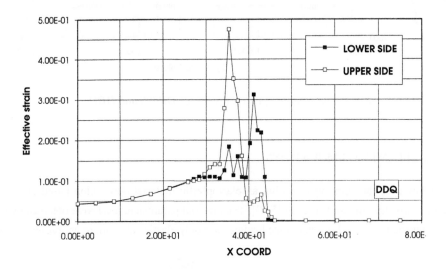

Fig. 7.   *Distribution of effective forming strains.*

Fig. 8 in turn shows the sheet thickness distribution along the x axis after forming. The two extreme values of the thickness correspond to the strain peaks shown in Fig. 9. For comparison an experimental value of the sheet thickness in the fatigue critical area is shown. The agreement between the numerical value and the experimental is good although the finite element simulation underestimates the thickness reduction.

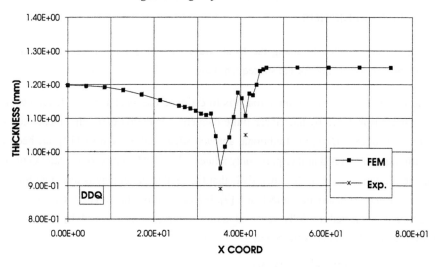

Fig. 8.   *Distribution of sheet thickness after forming simulation.*

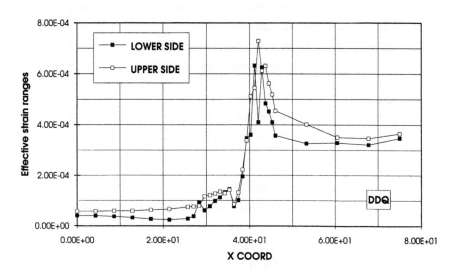

Fig. 9.   *Distribution of effective strain ranges during fatigue load simulation.*

Fig. 9 shows the effective strain range along the x axis of the sheet during fatigue loading. The peak strain range is higher on the upper side of the sheet than on the lower side. The peak occurres at the end portion of the coining. The largest peak is thus obtained in the location where crack were found to initiate in the experiments. The elastic strain concentration factor (=peak strain/nominal strain) was found to be 2.2.

4. FATIGUE LIFE PREDICTION

The fatigue life was predicted by using the local strain aproach for the fatigue crack initation life [1,12,15] and fracture mechanics for the subsequent fatigue crack growth [2,3].

The essential feature of the local strain aproach is to estimate the local strain in fatigue critical area. This can be done for instance by non-linear finite element calculation or by the application of strain gauges to the critical area. Here, the traditional aproach suggested by Neuber [12] has been adopted. The point with this analysis is that it requires only a linear elastic calculation of the stress concentration factor. A modern interpretation of this analysis is that the total strain energy is the same for the elastic case and the requested cyclically non-linear case [9]. For ease of use the fatigue life data obtained from smooth specimens subjected to fully reversed strain control is established by using the Neuber parameter (left hand side) and is expressed as

$$\sqrt{E \Delta \sigma \Delta \varepsilon} = D \cdot N_f^d \tag{2}$$

the fitting parameters D and d used were 5169 and -0.204 for DDQ and 4210 and -0.173 for DP400. The requested initation fatigue life was obtained by calculating the linear elastic peak strain by multiplying the nominal strain with the stress concentration factor of 2.2 established in the finite element simulations. The elastic peak strain and stress was inserted into the left hand side of Eq. (2).

Fig. 10 shows the predicted fatigue life vs the experimentally determined fatigue lives of Fig. 7. A perfect relation between predicted and experimental lives follow the solid diagonal in the figure. We can see that local strain aproach for prediction of the initiation fatigue life is conservative with about a factor of 5.

In addition to the fatigue life initiation estimate, a fatigue crack growth analysis was performed. This was performed by integrating Paris' law [14] from an initial crack length of 0.6 mm to a final crack length of 20 mm. The parameters in Paris' law was $C=3.63*10^{-13}$ and $m=4.1$ [16]. The parameters does not vary much for the two materials. The initial crack length chosen is close to the threshold crack length typical for sheet steels, $\Delta K_{th}=4.1\text{-}4.6$ MPa$\sqrt{m}$ [16]. The final crack length corresponds to the crack length observed at termination of the fatigue tests. The stress intensity solution for the crack was that for through the thickness crack in a plate subjected to pure bending [11]. Since the crack in the case of the coined geometry grows in the stress concentration area the bending stress of the stress intensity solution was multiplied by the stress concentration factor.

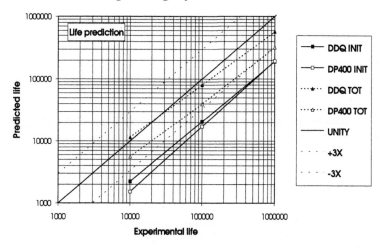

Fig. 10.   Evaluation of predicted fatigue lives using either the local strain
aproach for fatigue crack initiation life (INIT) or the local strain aproach plus
fatigue crack growth analysis for the total fatigue life (TOT).

In Fig. 10 also the predicted life obtained by superposing the crack growth life to the initiation life can
be seen. The predicted fatigue life is improved by adding the fatigue crack growth; within a factor of 3
in life.

5. CONCLUSIONS

From the present investigation on the fatigue behaviour and fatigue life prediction of coined sheet
components the following conclusions can be made

- There is an advantage to utilize higher strength sheet materials than conventional forming steels.
  Here an increase in fatigue life of about a factor of 4 was noticed when comparing a conventional
  deep-drawing steel (DDQ) to a high-strength dual-phase sheet steel (DP400).

- The presented finite element calculations were shown to produce a realistic forming simulation. It
  was shown that the calculated sheet thickness reduction in the fatigue critical area was close to the
  experimentally determined one.

- Fatigue life predictions based on fatigue life data for smooth specimens subjected to fully reversed
  strain controlled testing and local strain ranges from finite element calculations were shown to be
  overly conservative - about a factor of 5 in life. This prediction corresponds to fatigue crack
  initiation.

- A fatigue crack growth correction for the fatigue life estimation were able to explain the conservativeness of the crack initation estimate.

6. ACKNOWLEDGEMENTS

The following organizations are acknowledged for their financial support: NUTEK, SSAB Tunnplåt AB, Volvo Cars and Trucks.

7. REFERENCES

[1]   DOWLING, N, A Discussion of Methods for Estimating Fatgiue Life, (1982) Proc. SAE Fatigue Conf., P 109, Society of Automative Engineers, Warrendale, PA, pp 161-174.

[2]   EWALDS, H.L. and WANHILL, R.J.H., (1984), Fracture Mechanics, Edward Arnold/Delftse Uitgevers Maatschappij.

[3]   GUSTAVSSON, A.I., Variable Amplitude Fatigue of Notched Sheet Steels - An Overview of Mechanisms and Mechanical Models, (1992), Swedish Institute for Metals Research, IM-2869, Stockholm, Sweden.

[4]   GUSTAVSSON, A. and LARSSON, M., Fatigue Of Coined Sheet Steels. Part 1: Strain Controlled Fatigue Testing Of Different Low Carbon Sheet Steels, (1993), Swedish Institute for Metals Research, IM-2930, Stockholm, Sweden.

[5]   GUSTAVSSON, A., LARSSON, M. and MELANDER, A., New Techniques To Test Fatigue Properties Of Coined Sheet Specimens Part Three: Fatigue Testing And Analysis Of A Component Like Coining In Sheet Steel Materials, (1994), IM-3146, Swedish Institute for Metals Research, Stockholm, Sweden.

[6]   GUSTAVSSON, A. and MELANDER, A., New Techniques To Test Fatigue Properties Of Coined Sheet Specimens Part Four: Development Of Experimental Techniques, (1994), IM-3147, Swedish Institute for Metals Research, Stockholm, Sweden.

[7]   GUSTAVSSON, A. and MELANDER, A., Fatigue of a Highly Prestrained Dual-Phase Sheet Steel, (1995), Fatigue Fract. of Engng Mater. Struct. **18(2)**, pp 201-210.

[8]   HOLT, J.M. and CHARPENTIER, P.L., Effect of Cold Forming on the Strain-Controlled Fatigue Properties of HSLA Steel Sheets, (1983), Proc. Int Conf. on Technology and Applications of HSLA Steels, Philadelphia, pp 209-222.

[9]   MOFTAKHAR, A., BUCZYNSKI, A., and GLINKA, G., Calculation fo Elasto-Plastic Srains and Stresses in Notched Bodies under Multiaxial Cyclic Loading, (1993), Proc. 5th Int. Conf. Fatigue and Fracture, 3-7 May 1993, Montreial, Canada, EMAS, pp 441-452.

[10]   MOOSBRUGGER J C and MCDOWELL D L ,A rate dependent bounding surface model with a generalized image point for cyclic nonproportional viscoplasticity, (1990), J Mech Phys Solids **38**, 627-656.

[11]   MURAKAMI, Y., Stress intensity factor handbook, Vol. 2, (1987), Pergamon Press.

[12]   NEUBER, H., Theory of Stress Concentration for Shear-Strained Prismatical Bodies with Arbitrary Stress-Strain Law, (1961), J Appl Mech, Dec, pp 544-550.

[13] PARKER, T.E. and MONTGOMERY, G.L., Effect of Balanced Biaxial Stretching on the Low-Cycle Fatigue Behaviour of SAE 1008 Hot Rolled Low Carbon Steel, (1975), SAE Automotive Engineering Congress and Exposition, Paper No. 750048, Detroit.

[14] PARIS, P.C., GOMEZ, M.P. and ANDERSON, W.E., A Raional Analytic Theory of Fatigue, (1961), The Trend in Engineering, pp9-14.

[15] SHERRATT, F., and EATON, O., Fatigue Life Estimation by Local Stress-Strain Methods, (1983), J. Society of Env. Eng., Sept., pp 28-36.

[16] WASÉN, J., Fatigue Crack Growth and Fracture in Steel, (1988), thesis, Chalmers University of Technology, Department of Engineering Metals.

[17] WEI, D.C., Structure-Fatigue Corrections for Dual Phase HSLA Steels, (1981), SAMPE quarterly **12(4)**, pp 24-31.

A MULTIAXIAL FATIGUE LIFE PREDICTION PROGRAM

T. E. LANGLAIS

125 Mechanical Engineering, University of Minnesota, Minneapolis, MN, USA

J. H. VOGEL

SciMED Life Systems, 2010 E. Center Cirle, Plymouth, MN, USA

D. F. SOCIE

University of Illinois at Urbana-Champaign, 1206 West Green, Urbana, IL, USA

T. S. CORDES

Deere&Co. Technical Center, 3300 River Dr., Moline, IL, USA

ABSTRACT

A step-by-step process for multiaxial fatigue life prediction is outlined. The approach is built from existing models and methods, modified to fit into integrated software. The process begins with an elastic finite element analysis to determine geometry factors, continues with stress-strain calculation via notch correction/plasticity, and finishes with critical plane damage estimation to assess fatigue lives. The use of FEA and notch correction/plasticity to calculate local stress-strain behavior is discussed and critical plane methods are reviewed. As an example of the complete process, the integrated method is used to generate life predictions for the Society of Automotive Engineers (SAE) notched shaft from a sample load history.

KEYWORDS

Multiaxial Fatigue, Multiaxial Notch Correction, Multiaxial Plasticity, Critical Plane Analysis, SAE Shaft

NOMENCLATURE

$\underset{\sim}{s}, \underset{\sim}{\alpha}$ deviatoric stress tensor, deviatoric backstress tensor
$\underset{\sim}{\epsilon}, \underset{\sim}{e}$ strain tensor, deviatoric strain tensor
$\underset{\sim}{\epsilon}^p, \underset{\sim}{e}^p$ plastic strain tensor, non-linear part of the fictitious (elastically calculated) strain

85

$\bar{\sigma}, {}^e\bar{\sigma}$ equivalent stress, equivalent fictitious stress
$\bar{\epsilon}, {}^e\bar{\epsilon}$ equivalent strain, equivalent fictitious strain
$\bar{\epsilon}^p, {}^e\bar{\epsilon}^p$ equivalent plastic strain, non-linear part of the equivalent fictitious strain
$C, {}^eC$ generalized plastic modulus, fictitious generalized plastic modulus
D_t, D_s tensile, shear damage
σ_f', τ_f' tensile, shear fatigue strength coefficient
ϵ_f', γ_f' tensile, shear fatigue ductility coefficient
σ_{max} maximum stress *normal to* the plane

INTRODUCTION

Most engineers in the ground vehicle industries today rely on uniaxial analysis in the context of the local strain approach to predict fatigue life. These methods have proven to be effective in correlating fatigue life for many parts whose loading can be assumed uniaxial. However, the interest in multiaxial methods has grown as the ability to measure multiaxial loading histories and perform three dimensional stress analysis becomes more routine. Designers frequently ask: why do we do uniaxial fatigue analysis when we measure and compute multiaxial loads, stresses, and strains?

While researchers have demonstrated the ability of multiaxial methods to correlate crack nucleation life and direction [13], it is not clear that multiaxial methods, which come with a large computational price, are worth the added complexity. It is not the purpose of this study to definitively demonstrate the superiority of multiaxial methods; rather, the purpose is to show, by outlining a step-by-step process, how multiaxial fatigue life prediction can be performed on a real part and discuss the possible benefits of such an analysis.

The strength of multiaxial methods lies in their ability to account for behavior that is otherwise ignored when one performs a uniaxial analysis. There are cases—an axle is a good example— where the multiaxial state of stress is critical, loading is far from proportional, and crack initiation direction is important because subsequent loads may grow these cracks to failure. For these, a multiaxial analysis is appropriate.

Figure 1 shows the proposed multiaxial life prediction process schematically. Both notch correction and plasticity are used to calculate the stress-strain state in the notch. Several notch correction models have been developed for multiaxial stress-strain states [1,7,8,11]. Likewise, many researchers have proposed successful plasticity models for use in fatigue [4,9,10]. Critical plane methods, first developed by Findley [6] and later extended to strain by Brown and Miller [2], are used to estimate the fatigue life. Note that the method requires three major inputs: loads, geometry information in the form of elastic FEA, and material properties. Analysts can readily characterize the material properties using well-established uniaxial test procedures.

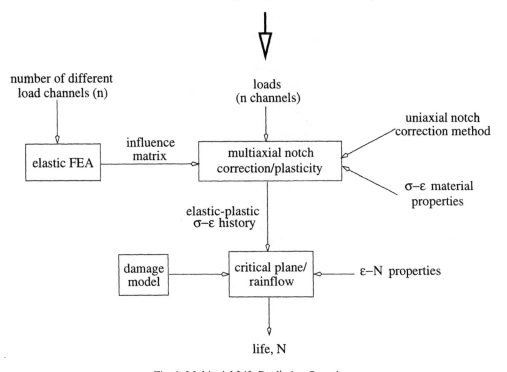

Fig. 1. Multiaxial Life Prediction Overview

METHODOLOGY

Influence Matrix

In order to begin a fatigue life analysis, something must be known about a part's service history; most often, this information comes in the form of load histories measured during simulations or during actual use. Although it could be experimentally determined or analytically derived, an influence matrix relating loads to in-plane elastic strains in the notch is most often found using elastic FEA,

$$ {}^e\epsilon_i = \underset{\sim}{L}\, P_j \tag{1} $$

where P_j is the vector of loads, $\underset{\sim}{L}$ is the influence matrix, and ${}^e\epsilon_i$ is a vector of elastically calculated in-plane notch strains. For the example shaft problem shown in Figure 2, the influence matrix is determined thus:

(i) Identify the number of different loads, usually corresponding to the number of load channels in a service history (e.g. $P = \{B\ T\}'$, where B is the bending moment and T is the torque)

(ii) Establish a suitable mesh in the elastic FEM software of choice

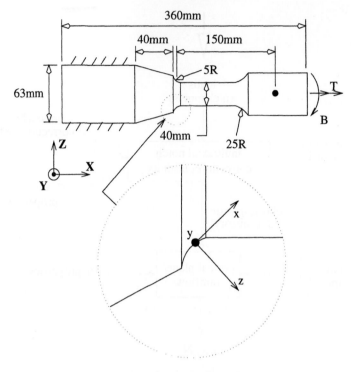

Fig. 2. SAE Notched Shaft

(iii) For each load type (channel), apply a singular unit load (two load cases for Figure 2, $P^B = \{1\,0\}'$ and $P^T = \{0\,1\}'$)

(iv) Solve for the elastic stresses and strains in the area of interest; these strains are the components of the influence matrix. For the bending moment

$$^e\epsilon_i = \left\{ \begin{array}{c} \epsilon^B_{xx} \\ \epsilon^B_{yy} \\ \epsilon^B_{xy} \end{array} \right\} \tag{2}$$

(v) By application of superposition, assemble the components into the final influence matrix

$$\left\{ \begin{array}{c} ^e\epsilon_{xx} \\ ^e\epsilon_{yy} \\ ^e\epsilon_{xy} \end{array} \right\} = \left[\begin{array}{cc} \epsilon^B_{xx} & \epsilon^T_{xx} \\ \epsilon^B_{yy} & \epsilon^T_{yy} \\ \epsilon^B_{xy} & \epsilon^T_{xy} \end{array} \right] \left\{ \begin{array}{c} B \\ T \end{array} \right\} \tag{3}$$

Using the influence matrix, it is easy to find the elastically calculated notch strains for any loading combination.

Notch Correction/Plasticity

Once the elastically calculated strains are known, a notch correction/plasticity model is used to find the elastic-plastic stress-strain state. Our approach to plasticity and notch correction is based on the following observations: 1) multiple-surface Mróz models do a good job estimating response to proportional loading [10], 2) two-surface models predict nonproportional material response accurately [9], and 3) correction models used to find notch strains usually employ a plasticity model in stress control [1]. The first two observations led to the choice of a modified Mróz-type plasticity model with a limit surface. The third led to the incorporation of Köttgen's $\underset{\sim}{\epsilon}$ notch correction method [8] into the plasticity model.

Loading conditions are divided into two areas: proportional and nonproportional. Proportional loading is best characterized by kinematic hardening whereas nonproportional material response, especially neutral loading, can be described by limit behavior. Here we use Chu's Mróz plasticity model [3] for kinematic hardening effects. For cases where the input equivalent strains do not change, the equivalent stress should not grow when the model assumes stabilized behavior. The Mróz model [10] has some difficulty dealing with this loading scenario; a limit surface is used to constrain the equivalent stresses when loading is nonproportional. The limit surface size is chosen according to the largest equivalent strain in a history.

Each increment of input elastically calculated strain, $d^e\underset{\sim}{\epsilon}$, is broken into a two-step process that amounts to simultaneous numerical integration. The process is detailed below.

A: $\cdot\, d\underset{\sim}{\overset{e}{\epsilon}} \Rightarrow d\underset{\sim}{s}$

     $\cdot$ use $d\underset{\sim}{\overset{e}{\epsilon}}{}^P = \frac{1}{{}^eC}\, g\left(\underset{\sim}{s} - \underset{\sim}{\alpha}\right)$, $\Delta^e\underset{\sim}{e} = \frac{\underset{\sim}{s}}{2G}$, ${}^eC = \frac{2}{3}\frac{d\bar{\sigma}}{d^e\bar{\epsilon}^P}$

     $\cdot$ numerically integrate $d\underset{\sim}{s} = f_s\left({}^eC, \underset{\sim}{s}, \underset{\sim}{\alpha}\right)$

B: $\cdot\, d\underset{\sim}{s} \Rightarrow d\underset{\sim}{\epsilon}$

     $\cdot$ use $d\underset{\sim}{\epsilon}^P = \frac{1}{C}\, g\left(\underset{\sim}{s} - \underset{\sim}{\alpha}\right)$, $\Delta\underset{\sim}{e} = \frac{\underset{\sim}{s}}{2G}$, $C = \frac{2}{3}\frac{d\bar{\sigma}}{d\bar{\epsilon}^P}$

     $\cdot$ numerically integrate $d\underset{\sim}{\epsilon} = f_e\left(C, \underset{\sim}{s}, \underset{\sim}{\alpha}\right)$

Critical Plane Analysis

The critical plane approach tries to do analytically what the test engineer does experimentally. To analyze a failure, a test engineer finds the location of crack nucleation and places a strain gauge at that location in a direction perpendicular to the observed crack and measures the strain history. Analytically, the location and direction are unknown and all possiblities must be evaluated. It is assumed that the plane and direction having the most computed damage will be the first one to nucleate a crack.

As yet, there is no consensus among researchers as to which damage model is best and this paper will not add to this debate. Nevertheless, we may observe that successful models have features that

- account for plasticity using a strain-based parameter
- account for nonproportional hardening using a stress range or maximum stress parameter
- model mean stress effects with a mean stress or maximum stress parameter
- account for different failure modes by characterizing damage as tensile or shear related

No single method meets all of these requirements, the most difficult being the ability to distinguish between tensile and shear damage modes.

Shear damage is estimated using

$$D_s = \frac{1}{N_s} \tag{4}$$

where N_s is found from

$$\frac{\Delta\gamma}{2}\left(1 + k\frac{\sigma_{max}}{\sigma'_f}\right) = \frac{\tau'_f}{G}(2N_s)^b + \gamma'_f(2N_s)^c \tag{5}$$

This model was first proposed by Fatemi and Socie [5]. The right-hand side is the strain-life curve generated from torsion testing. The terms on the left-hand side represent the loading parameters defined on a given plane.

The Smith-Watson-Topper (SWT) model [12] is used to compute the tensile damage:

$$D_t = \frac{1}{N_t} \tag{6}$$

where N_t is found by solving

$$\sigma_{max}\frac{\Delta\epsilon}{2} = \frac{\Delta\sigma_r}{2}\left(\frac{\sigma'_f}{E}(2N_t)^b + \epsilon'_f(2N_t)^c\right) \tag{7}$$

The right-hand side of the equation is the SWT parameter ($\sigma_{max}\Delta\epsilon/2$) for fully reversed uniaxial loading case in terms of the strain-life curve. The left-hand side of the equation represents the loading parameters. In this paper we have used a simple but effective strategy: compute shear and tensile damage on each plane and add the damage together to obtain an estimate of component life, N:

$$\frac{1}{N} = \frac{1}{N_t} + \frac{1}{N_s} \tag{8}$$

Because the tensile and shear cracking modes do not often occur simultaneously, summing the tensile and shear damage leads to a conservative estimate for the fatigue life.

Table 1. Comparison of Uniaxial and Multiaxial Notch Correction Procedures

	nominal strain (μe)	fictitious strain (μe)	notch strain (μe)	notch stress (MPa)
$K_f = 1.0$				
uniaxial	2690	2690	3190	467
multiaxial	2690	2690	3130	476
$K_f = 2.0$				
uniaxial	2690	5380	9030	660
multiaxial	2690	5380	8945	697

Summary of Methodology

In summary, the integrated approach takes this form:

(i) Identify the regions of interest and the inputs
(ii) Determine the influence matrix with use of elastic FEA
(iii) Feed the notch correction/plasticity model the separate channels of the load history to find the appropriate stress-strain states
(iv) Rainflow count the resulting stress-strain history, resolved on a plane, using the damage model of choice
(v) Calculate the final life by identifying the most damaged plane

COMPARISON TO EXPERIMENT

It is important to verify that the multiaxial stress-strain predictions agree with the uniaxial methods on which they are based. Shown in Table 1 are the results of simple uniaxial loading with notch correction. Note that the multiaxial method reduces to expected behavior in uniaxial tension. The comparison is made between a uniaxial Neuber's analysis and the integrated notch correction/plasticity model for SAE 1045 steel, hardened to Rockwell C 29.

Figure 3 represents computed and experimental data for a simple tension ($S_{xx} = 296$MPa) and torsion ($S_{xy} = 193$MPa) loading of a circumferential notched shaft. The nominal quantities are computed from the applied torque, T, and the applied axial load P as $S_{xx} = P/A$ and $S_{xy} = TR/J$, where A is the cross-sectional area, R is the nominal radius, and J is the rotational moment of inertia. Note that the coupling between the shear and axial loading is captured by the model. The notch shear strain increases during the tensile loading portion of the loading cycle even though the nominal shear stress is held constant during this loading segment.

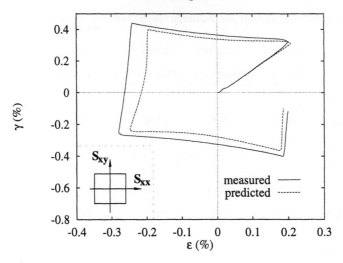

Fig. 3. Box Path, Notch Correction

SAMPLE ANALYSIS OF SAE NOTCHED SHAFT

The SAE notched shaft is shown in Figure 2. The step-by-step analysis procedure will be detailed here by example. Step numbers refer to the summary procedure in Section *Summary of Methodology.*

STEP 1
The shaft is loaded by torsion and one plane of bending. These represent the two load channels for this case, $P = \{B\ T\}'$. A mesh is chosen with sufficient refinement in the area of the notch to determine the location with the largest strains.

STEP 2
An elastic FEA for two load cases—one with a 1000Nm bending load and no torque, the other with a 1000Nm torsion load and no bending—locates the largest strain approximately 15° up in the notch root for both load cases. In the rotated local coordinate system shown in Figure 2, the resulting strains are:

$$\begin{Bmatrix} \epsilon_{xx}^{B} \\ \epsilon_{yy}^{B} \\ \epsilon_{xy}^{B} \end{Bmatrix} = \begin{Bmatrix} 1176 \\ -128 \\ 0 \end{Bmatrix} \mu e \quad \begin{Bmatrix} \epsilon_{xx}^{T} \\ \epsilon_{yy}^{T} \\ \epsilon_{xy}^{T} \end{Bmatrix} = \begin{Bmatrix} 0 \\ 0 \\ 641 \end{Bmatrix} \mu e$$

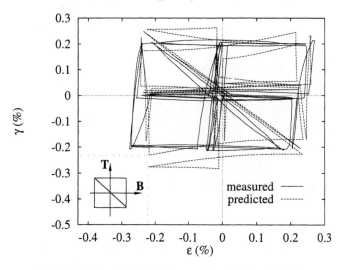

Fig. 4. SAE Notched Shaft Measured and Predicted Strains

The influence matrix for a $1Nm$ load can be assembled:

$$\underset{\sim}{L} = \begin{bmatrix} 1.176 & 0.0 \\ -0.128 & 0.0 \\ 0.0 & 0.641 \end{bmatrix}$$

STEP 3

Figure 4 shows the predicted and measured notch strains for the SAE shaft. An elastic test was used to determine an experimental influence matrix because of the inability to place the gauge at the location of highest strain predicted by FEA and thereby make a fair comparison.

STEP 4 & STEP 5

The resulting damage map is shown in Figure 5, where ϕ and θ locate the normal to the plane, n, as shown (the z-axis is normal to the surface of the shaft as in Figure 2) and D is the fatigue damage. The method estimates a fatigue life of 65 blocks on the critical plane $\phi = 170^{\circ}, \theta = 90^{\circ}$. Each block consists of ten repetitions of the load history.

DISCUSSION

The advantage of the multiaxial method lies in its ability to describe nonproportional loading while reducing to expected behavior in proportional loading. Our particular method combines into one step the solution of the notch strains and stresses, unlike many proposed methods that separate the two. This integration saves computation time.

The loading history used in this example turns out to be a very challenging test of the notch

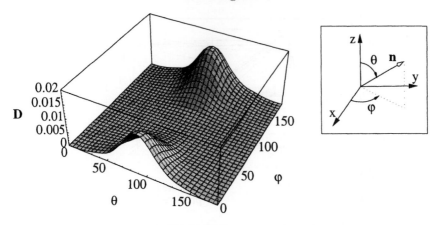

Fig. 5. Predicted Damage Distribution for Nonproportional Loading of SAE Shaft

correction/plasticity methodology. It contains mostly out-of-phase loading, while our approach was developed with an emphasis on accurately matching response to uniaxial and in-phase loading. This is, in some sense, a worst case analysis, although an important one, since the history contains many features common to loadings on components like axles.

The uniaxial properties we used for this analysis are based on the work of Fatemi [5], who reported stress-strain response for tubes loaded in out-of-phase torsion and axial tension/compression. The cyclically stabilized axial stress vs. axial strain corresponding to zero torsion was fit to give $k' = 3644$ MPa and $n' = 0.3112$, with a modulus of 206519 MPa.

As shown in Figure 4, our estimates of the notch strains show somewhat more ratcheting behavior than is actually observed. The comparison is shown for the second highest load level tested ($T = 1795Nm$, $B = 1735Nm$) because there was evidence of strain gage de-bonding when the final highest load level test ($T = 2200Nm$, $B = 2100Nm$) was completed. This highest load level was run until part failure.

The predicted life of 65 blocks compares favorably to the measured life of 102 blocks. The experimental life corresponds to complete failure of the specimen into two pieces; the actual life to crack initiation is closer to the 65 block prediction.

CONCLUSIONS

An integrated procedure for multiaxial fatigue analysis has been outlined. The notch correction/plasticity model has been shown to qualitatively capture material behavior and the total process from input loads to final life estimation detailed for the SAE notched shaft. Life estimates within a factor of two were obtained for the notched shaft subjected to nonproportional loading.

REFERENCES

[1] Barkey, M. E., Socie, D. F., and Hsia, K. J. (Trans ASME 116, 1994). *Journal of Engineering Materials and Technology*, pp. 173–180.

[2] Brown, M. W. and Miller, K. J. (1973). In:*Proceedings Institute of Mechanical Engineers*, **187**, 65, pp. 745–755.

[3] Chu, C. C. (1992). SAE Technical Paper Series 920662, Society of Automotive Engineers, Warrendale, PA.

[4] Dafalias, Y. F. and Popov, E. P. (1975). **23** *Acta Mechanica*, pp. 173–192.

[5] Fatemi, A. and Socie, D. F. (1988). *Fatigue and Fracture of Engineering Materials and Structures*, **11**, pp. 149–165.

[6] Findley, W. N., Coleman, J. J., and Handley, B. C. (1956). In: *International Conference on Fatigue of Metals*, The Institution of Mechanical Engineers, New York, pp. 150–157

[7] Hoffman, M. and Seeger, T. (1989). In: *Multiaxial Fatigue:Analysis and Experiments, AE-14*, G. E. Leese and D. F. Socie (Ed.). Society of Automotive Engineers, Warrendale, PA, pp. 81–99.

[8] Köttgen, V. B., Barkey, M. E., and Socie, D. F. (1995). *Fatigue and Fracture of Engineering Materials and Structures*, **18**, 9, pp. 981–1006.

[9] McDowell, D. L. (1985). *Journal of Applied Mechanics*, **52**, pp. 298–308.

[10] Mróz, Z. (1967) *Journal of the Mechanics and Physics of Solids*, **15** pp. 163–175.

[11] M. N. K. Singh, G. Glinka, and R. N. Dubey (1997) In: *Proceedings of the 1995 Symposium on Multiaxial Fatigue and Deformation Testing Techniques*, SAE 1280, pp. 131–135.

[12] Smith, R. N., Watson, P., and Topper, T. H. (1970). *Journal of Materials*, **5**, 4, pp. 767–778.

[13] Socie, Darrell (1993). In: *Advances in Multiaxial Fatigue, ASTM STP 1191*, D. L. McDowell and R. Ellis (Eds.).American Society for Testing and Materials, Philadelphia, pp. 7–36.

FATIGUE FAILURE OF A CONNECTING ROD

ROGER RABB

Wartsila Diesel International Ltd, PO Box 244, FIN - 65101, Vaasa, Finland

(*Received* 12 *September* 1995)

Abstract—This paper describes the analysis of a fatigue failure of a connecting rod in a medium-speed diesel engine. The difficulties in making a sufficiently good FE model with exact geometry of fine details and with all important nonlinearities are explained. Fatigue tests of the material in the connecting rod were also carried out. The FE analyses and fatigue data led to an improved design of connecting rod.

1. BACKGROUND

In 1989, an 18-cylinder diesel engine experienced a connecting rod failure at a steel mill. The purpose of the engine was to run an electric generator that provided the electric furnaces with current. The consequences were disastrous: the engine block was smashed in two and the crankshaft was ruined, i.e. the whole engine was destroyed. Fortunately nobody was hurt, but the economic consequences were serious. The connecting rod design is shown in Fig. 1.

An examination of the fracture surfaces (Figs 2 and 3) revealed that the cause of the failure was fatigue and that the crack had initiated in the upper connecting rod .

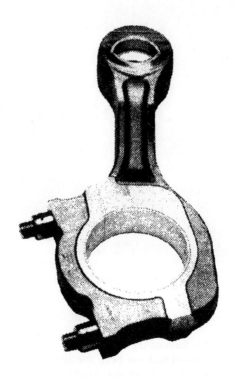

Fig. 1. Connecting rod.

Fig. 2. Broken shaft of the failed connecting rod.

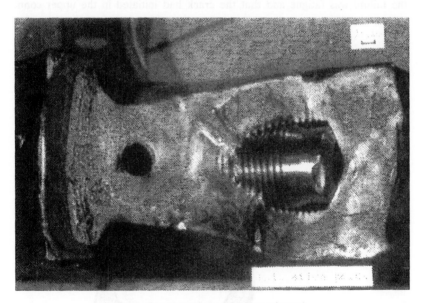

Fig. 3. Fracture surface of the connecting rod.

threads. The initiation point was in the root of the fifth thread from the end of the engagement.

A measurement of the form of the thread profile showed that the root radius of the thread in the connecting rod had not been specified, as could be concluded from the working drawings. The standards do not require any root radius for the thread of a nut. The result was that the root radius of the thread in the connecting rod varied from about 0.1 mm to about 0.2 mm. The root radius in the connecting rod screw was, in accordance with the appropriate standard, equal to 0.3 mm.

It was obvious that the stresses in the connecting rod in the threaded area had to be checked with a demanding FE analysis. The calculations made at the design stage had not been exact enough to calculate the stresses in the zone where the crack had initiated. Although we had a powerful FE program (ADINA) our computer resources were a limiting factor. In a nonlinear analysis it was necessary to restrict the size of the model to about 20,000 DOFs, which was insufficient. We therefore had to look to a consultant for increased computing power. The FE model would have to involve nonlinearities like contact between the flanks of the threads, and details such as the threads would have to be modelled accurately. The consultant chosen had a Convex C1 minisuper computer with the well-known FE program ABAQUS.

2. CHOICE OF GOALS FOR THE ANALYSIS

The size of FE model needed would have reached the limits of the consultant's computer. It was, therefore, necessary to simplify the model and its boundary conditions, and to consider whether it was really important to include a given nonlinearity in the model. In the first analysis commissioned it was decided that the only nonlinearity to be included would be contact without friction. As the first results had shown that considerable slip occurred between some engaged flanks, it was decided that friction had to be included. When the results of the first fatigue analysis revealed large failure probabilities, it was clear that the load equalizing influence of thread root plasticity had to be included in the model. Finally, with plasticity and friction in the model, it was necessary to give load cases in sequence and to repeat the load sequences to reach the steady-state situation.

The following were established for the original analysis:

(a) Load cases (not given in sequence).
● Application of prestress in the connecting rod screws.
● Application of a diametrical interference fit between the bearing sleeve and the big end of the connecting rod together with the prestress in the screws.
● Application of inertia forces together with the action from the interference fit and the screw prestress.
● Application of the combined effect of maximum combustion pressure and the inertia forces together with the interference fit and the screw prestress.
(b) Boundary conditions.
● The supporting effect of the crank pin was correctly taken into account as a contact problem by modelling the pin with the appropriate diametrical clearance. Friction effects were neglected at this location.

The aim of this first analysis was to see the differences in stress amplitudes and mean stresses with a root radius in the connecting rod thread of 0.15 mm and with an appropriately specified root radius of 0.3 mm. It was assumed that a linear material model would suffice for this comparison. Our experience of failure analysis was based on the use of a nominal stress and a corresponding stress concentration factor. This approach is not well suited for FE analyses, where real stresses are calculated and the definition of a nominal stress is difficult.

3. THE FIRST FE MODEL

3.1. *An axisymmetric FE analysis*

The benefit of an axisymmetric FE model is that it is easy to modify and run. It is also easier to agree upon some nominal stress. The axisymmetric FE model is shown in Figs 4 and 5. Figure 6 shows the effective stress distribution in some of the first threads in engagement when a force acts at the end of the screw shank. Initially, it was decided to use the stress concentration factors obtained with the axisymmetric model for a tension force through the outer cylindrical part. This is a rather conservative estimate, although it was considered accurate enough to allow comparison between different alternatives. The screw dimension was M45 × 3 and the

R. Rabb

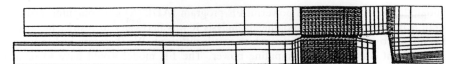

Fig. 4. Axisymmetric FE model to determine stress concentration factor.

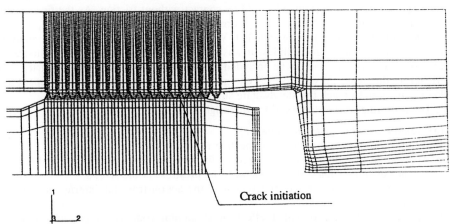

Fig. 5. Threads of the axisymmetric FE model.

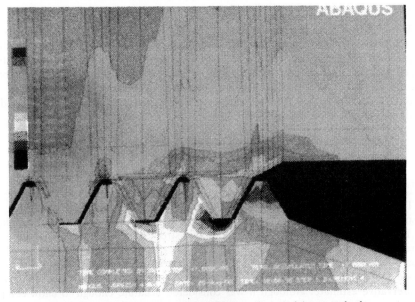

Fig. 6. Effective stresses when a force is acting at the end of the screw shank.

calculated stress concentration factors were $K_t = 3.7$, for a bottom radius of 0.15 mm, and $K_t = 2.7$, for a bottom radius of 0.3 mm.

As expected, the ratio of these two stress concentration factors is approximately equal to the square root of the inverse of the ratio of the bottom radii. The results also showed that the maximum principal stress was dominant. Hence, the fatigue analysis could, with sufficient accuracy, be based on a uniaxial stress assumption and maximum principal stress at the surface.

3.2. *Global FE analysis*

Because of the complexity of the analysis, it was conducted in two steps. The first step was to create a global model of the entire connecting rod with linear 3D elements for modelling even the fine details of the threads. The global model was built of the following substructures:

- Substructure z101. This substructure contains the crack pin modelled with 3D elements with quadratic interpolation formulation.
- Substructure z102. This substructure contains the bearing of the big end of the connecting rod. It is modelled with 3D elements with quadratic interpolation formulation.
- Substructure z103. This substructure contains the connecting rod screw modelled with 3D elements with linear interpolation formulation.
- Substructure z104. This substructure contains the connecting rod modelled with 3D elements with quadratic interpolation formulation everywhere except for a cylindrical zone with an outer radius of 30 mm which contains the threads of the connecting rod. In this cylindrical area 3D elements with linear interpolation formulation are used.

Thus, the global model consists mainly of 20-node isoparametric 3D elements. The model takes into account the clearance between the crank pin and the connecting rod bearing. This global FE model is shown in Figs 7 and 8. Nodes on the nonconformical parts of substructure z104 are connected to the rest of the connecting rod with constraint equations.

The second step was to create a local FE model of the threaded area of the connecting rod and the connecting rod screw by using substructures z103 and z104, but now there were 3D elements with quadratic interpolation formulation in the substructure z104 of the threaded region. This local model was fine enough to allow an exact modelling of the threads (Figs 9 and 10). During the first runs, the engagement of the threads of the connecting rod and the screw was treated as a nonlinear contact problem without friction. Later on, the friction between the flanks was included in the model and later plasticity and load cycling were included as well. The load applied to this model was the displacement load that was found interpolated with the global model and the tension in the shaft of the screw. The big difference in the size of the elements of the global model and the local model naturally introduced accuracy problems when the global model displacements were interpolated to the locations of the corresponding nodes on the "outer" surface of the local model.

In ABAQUS one can choose either a total Lagrangian or a penalty function approach to solve a nonlinear contact problem. The penalty function approach was chosen because this algorithm converges faster. The method has a disadvantage in having to define the elastic slip if friction is involved. Making a suitable choice of elastic slip is difficult and can introduce some arbitrariness into the analysis. On a physical level one can assume that the elastic slip corresponds to elastic displacements in the surface roughness. The wrong choice of elastic slip can considerably influence the results of the analysis.

The purpose of the first runs was to examine how much a specified radius of 0.3 mm in the thread bottom of the connecting rod would increase the safety factors in comparison to the original design of a bottom radius of 0.15 mm. Tables 1 to 4

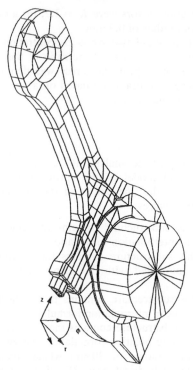

Fig. 7. Nonlinear global FE model of the connecting rod with a local cordinate system for the threads.

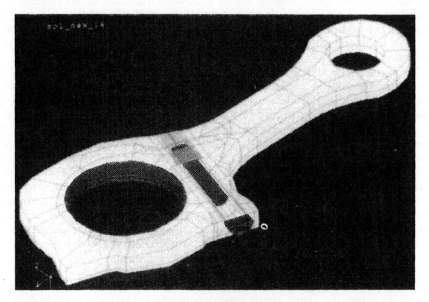

Fig. 8. Global FE model.

show the maximum principal stresses and nominal stresses in the root of the threads of the connecting rod and in the direction of the small end of the rod. The corresponding mean stresses and stress amplitudes are also shown. The type of steel used is forged 34CrMo4 TQ+T ISO 683-1, with a minimum tensile strength, R_m, of 750 MPa, and a minimum yield strength, $R_{p0.2}$, of 500 MPa.

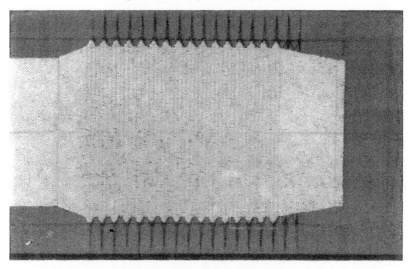

Fig. 9. Local model of the threads made from the substructures z103 and z104.

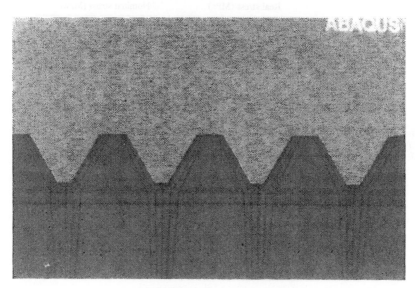

Fig. 10. View from Fig. 9. The thread mesh.

The calculations give rise to the following comments:

- The negative influence of a thread without bottom radius is clearly shown. The calculated stresses decrease considerably if a bottom radius of 0.3 mm is used.
- The yield stress is reached in many threads, especially in the beginning of the engagement. The load distribution between the threads is, of course, influenced considerably by this yielding, and an elastoplastic material model will be used in subsequent analyses.
- The calculated slips between thread flanks in engagement are very high when friction is neglected (52–58 μm). A microscopic study of the flanks revealed the truth (Fig. 11): the flanks had been damaged by fretting. It was therefore possible that the first crack was initiated by fretting rather than by an excessive stress amplitude. It was therefore necessary to bring friction into the analysis.

R. Rabb

Table 1. Original design (0.15 mm radius)

Flank no.	Peak stress (MPa)		
	Prestress	Prestress and inertia forces	Prestress and combustion
1	2904	2869	3188
2	1161	1156	1535
3	991	987	1354
4	890	886	1250
5	809	804	1165
6	728	722	1091
7	677	668	1040
8	629	619	996
9	579	567	945
10	538	525	913
11	531	515	912
12	534	516	940
13	645	624	1131
14	1311	1278	2245

Table 2. The original design (0.15 mm radius)

Flank no.	Real stress (MPa)		Nominal stress (MPa)	
	σ_m	σ_a	$S_m = \dfrac{\sigma_m}{3.7}$	$S_a = \dfrac{\sigma_a}{3.7}$
1	3028.5	159.5	818.5 > 500	43.11
2	1345.5	189.5	363.6	51.22
3	1170.5	183.5	316.4	49.69
4	1068.0	182.0	288.6	45.19
5	984.5	180.5	266.1	48.78
6	906.5	184.5	245.0	49.86
7	854.0	186.0	230.8	50.27
8	807.5	188.5	218.2	50.95
9	756.0	189.0	204.3	51.08
10	719.0	194.0	194.3	52.43
11	713.5	198.5	192.8	53.65
12	728.0	212.0	196.8	57.30
13	877.5	253.5	237.2	68.51
14	1761.5	483.5	476.1	130.68

Table 3. Modified design (0.3 mm radius)

Flank no	Peak stress (MPa)		
	Prestress	Prestress and inertia forces	Prestress and combustion
1	1260	1258	1527
2	893	895	1098
3	747	749	944
4	667	669	864
5	614	615	812
6	561	561	762
7	522	527	733
8	475	477	686
9	440	442	656
10	405	406	623
11	378	377	599
12	350	348	572
13	327	321	550
14	333	325	568
15	364	354	638

Table 4. Modified design (0.3 mm radius)

Flank no.	Real stress (MPa)		Nominal stress (MPa)	
	σ_m	σ_a	$S_m = \dfrac{\sigma_m}{2.7}$	$S_a = \dfrac{\sigma_a}{2.7}$
1	1392.5	134.5	515.7 > 500	49.81
2	995.5	102.6	368.7	37.96
3	845.5	98.5	313.1	36.48
4	765.5	98.5	283.5	36.48
5	713.0	99.0	264.1	36.67
6	661.5	100.5	245.0	37.22
7	627.5	105.5	232.4	39.07
8	580.5	105.5	215.0	39.07
9	548.0	108.0	203.0	40.00
10	514.0	109.0	190.4	40.37
11	488.0	111.0	180.7	41.11
12	460.0	112.0	170.4	41.48
13	435.5	114.0	161.3	42.41
14	446.5	121.5	165.4	45.00
15	496.0	142.0	183.7	52.59

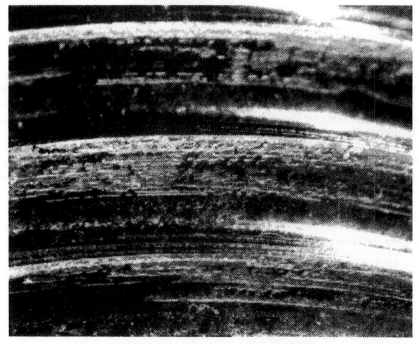

Fig. 11. Fretting wear damage on the thread flanks.

4. FATIGUE TESTING OF 34CrMo4 TQ+T STEEL

The fatigue testing of the steel of the connecting rod had several aims:

- To determine the ratio of the fatigue limit in alternating tension/compression for a smooth test bar to the ultimate tensile stress.
- To determine the fatigue limit of a notched test bar both with alternating tension/compression and with a nominal mean stress equal to the yield stress.
- To find the notch sensitivity of this steel.
- To find if there were any signs of material anisotropy (Fig. 12).

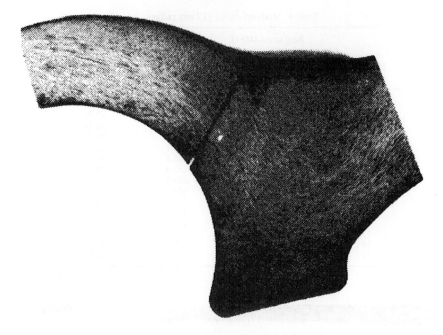

Fig. 12. Material grain flow in the area of the threads.

According to the quality instruction for drop forging of the connecting rod, the steel to be used can also be 34CrNiMo6 TQ+T EN 10083-1. In fact, most of the fatigue testing was done on this steel. Test bars were cut from the forging. For testing the anisotropy, they were cut in the same direction as the grain flow and perpendicular to it. According to the quality instruction steel 34CrNiMo6 has a tensile strength of 850–1000 MPa and a minimum yield strength of 700 MPa. Test results on material used for the tests gave mean tensile and yield strength values of 904 and 757 MPa, respectively.

4.1. *Testing for anisotropy*

These tests were carried out on smooth test bars made from 34CrNiMo4 TQ+T ISO 683-1. The results of the tests are summarized in Fig. 13.

As can be seen from Fig. 13, there is no clear indication of material anisotropy.

4.2. *Fatigue limit for an unnotched bar*

A small number of bars was used to examine the fatigue limit in alternating tension compression for an unnotched test bar. The results are shown in Fig. 14. The mean fatigue limit is given by

$$S_{C50} = S_{a0} + d \cdot \frac{A}{F} = 470 + 30 \cdot \frac{6}{5} = 506 \text{ MPa}. \tag{1}$$

The standard deviation is given by

$$s_a = d \cdot \sqrt{\frac{F \cdot B - A^2}{(F - 1) \cdot F}} = 30 \cdot \sqrt{\frac{5 \cdot 10 - 6^2}{(5 - 1) \cdot 5}} = 25 \text{ MPa}. \tag{2}$$

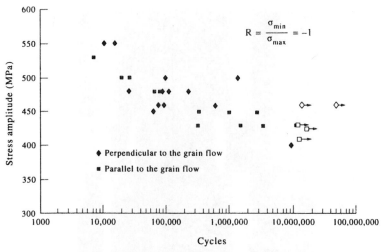

Fig. 13. Material anisotropy in drop-forged 34CrMo4 TQ+T.

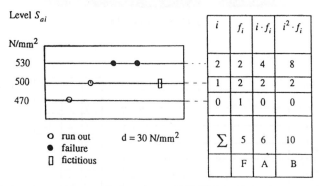

Fig. 14. Staircase test with unnotched test bars of 34CrNiMo6. Alternating tension/compression.

4.3. *Fatigue limit for a notched test bar*

The notch radius was 0.3 mm as in the root of the modified thread. The theoretical stress concentration factor was 3.63. Figure 15 shows a typical specimen. The fatigue limit was determined for both alternating tension/compression and for a (nominal) mean stress of 750 MPa, which is slightly higher than the minimum required yield stress of 700 MPa. For alternating tension compression with the notched bar we obtained the results in Fig. 16. The mean fatigue limit is given by

$$S_{C50} = S_{a0} + d \cdot \frac{A}{F} = 135 + 5 \cdot \frac{53}{18} = 150 \text{ MPa}. \tag{3}$$

The standard deviation is given by

$$s_a = d \cdot \sqrt{\frac{F \cdot B - A^2}{(F - 1) \cdot F}} = 5 \cdot \sqrt{\frac{18 \cdot 225 - 53^2}{17 \cdot 18}} = 10 \text{ MPa}. \tag{4}$$

Figure 17 gives the results for a (nominal) mean stress of 750 MPa in the notched bar. The mean fatigue limit is given by

$$S_{C50} = S_{a0} + d \cdot \frac{A}{F} = 70 + 3 \cdot \frac{13}{10} = 74 \text{ MPa}. \tag{5}$$

The standard deviation is given by

R. Rabb

Fig. 15. Notched test bar.

Level S_{ai}

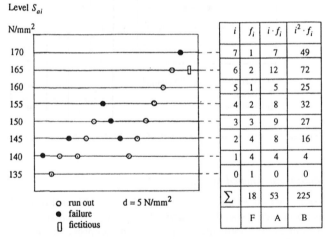

i	f_i	$i \cdot f_i$	$i^2 \cdot f_i$
7	1	7	49
6	2	12	72
5	1	5	25
4	2	8	32
3	3	9	27
2	4	8	16
1	4	4	4
0	1	0	0
Σ	18	53	225
	F	A	B

o run out $d = 5$ N/mm^2
● failure
▯ fictitious

Fig. 16. Test with notched test bars of 34CrNiMo6. Alternating tension/compression.

$$s_a = d \cdot \sqrt{\frac{FB - A^2}{F - 1) \cdot F}} = 3 \cdot \sqrt{\frac{10 \cdot 25 - 13^2}{9 \cdot 10}} = 3 \text{ MPa}. \tag{6}$$

4.4. Summary of fatigue results

The ratio of the mean fatigue limit in alternating tension/compression to the ultimate tensile stress is estimated as approximately

$$\frac{S_{C50}}{R_m} = \frac{506}{904} = 0.56. \tag{7}$$

The notch fatigue factor is given by

$$K_f = \frac{S_{C50K_t} = 1}{S_{C50K_t} = 3.63} = \frac{506}{150} = 3.38. \tag{8}$$

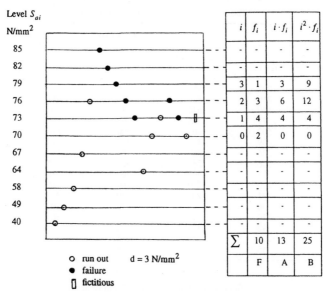

Fig. 17. Staircase test with notched test bars of 34CrNiMo6. Nominal mean stress $S_m = 750$ MPa.

The notch sensitivity factor is given by

$$\eta = \frac{K_f - 1}{K_t - 1} = \frac{3.38 - 1}{3.63 - 1} = 0.905.\tag{9}$$

Figure 17 shows that the use of Gerber's parabola and the rules for constructing a Haigh diagram give close agreement with the results of the tests. The value for the fatigue limit at a nominal mean stress of 750 MPa is 74 MPa, according to the Haigh diagram in Fig. 18, while the value found from the fatigue tests is also 74 MPa. Thus, the agreement is good, and we have obtained the somewhat surprising result that even with a nominal mean stress equal to the yield stress, our structure can be subjected to a considerable cyclic load.

5. DEVELOPMENT OF THE FE ANALYSIS

Friction and plasticity were introduced in later runs and many different alternatives were examined. The FE analysis yielded principal stresses shown in Table 5. The corresponding mean stresses and stress amplitudes are given in Table 6. A comparison with Tables 3 and 4 reveals the importance of introducing plasticity and strain hardening. There are no longer principal stresses far beyond the yield stress.

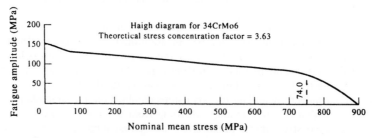

Fig. 18. Haigh diagram constructed with the aid of Gerber's parabola.

Table 5. Modified design (0.3 mm radius)

Flank no.	Prestress	Peak stress (MPa)			
		Prestress and inertia forces	Unloaded	Prestress and combustion	Unloaded
1	789	789	785	725	810
2	719	720	714	643	732
3	640	644	637	600	689
4	571	575	568	577	657
5	502	507	500	544	563
6	439	443	437	513	490
7	386	389	384	487	430
8	336	339	335	454	380
9	301	308	300	434	334
10	270	279	269	411	298
11	241	248	240	385	266
12	230	233	230	380	255
13	210	211	209	364	235
14	198	197	198	371	222
15	225	219	225	432	249
16	416	405	416	671	329

Table 6. Modified design (0.3 mm radius)

Flank no.	Real stress (MPa)		Nominal stress (MPa)	
	σ_m	σ_a	$S_m = \dfrac{\sigma_m}{2.7}$	$S_a = \dfrac{\sigma_a}{2.7}$
1	767.5	42.5	284.3	15.74
2	687.5	47.5	254.6	17.59
3	644.5	48.0	238.7	17.78
4	617.0	48.5	228.5	17.96
5	553.5	9.5	205.0	3.52
6	501.5	11.5	185.7	4.26
7	458.5	28.5	169.8	10.56
8	417.0	37.0	154.4	13.70
9	384.0	50.0	142.2	18.52
10	354.5	56.5	131.3	20.93
11	325.5	59.5	120.6	22.04
12	317.5	62.5	117.6	23.15
13	299.5	64.5	110.9	23.89
14	296.5	74.5	109.8	27.59
15	340.5	94.4	126.1	35.00
16	500.0	176.5	185.2	65.37

Figure 19 clearly shows the negative effect of too few load steps. The steady-state situation has not been reached and it remains unclear how to calculate the mean stress and the stress amplitude.

6. APPLICATION OF ANALYSIS WITH LOAD CYCLING

Load cycling was introduced into the analyses when a similar connecting rod was designed for a new engine. The number of additional load steps needed to achieve steady state was checked. Some results of this calculation are given below. Friction and plasticity were included. Figure 20 shows that steps 3 and 7 give the same stress level for the inertia load. In other words, we can conclude that the steady state has already been achieved by adding the load steps combustion and unloading to the load sequence. Note that due to friction, unloading from combustion and from inertia do not give the same stress level. Mean stress and stress amplitude in the steady-state situation is determined from load steps 5–7 as can be concluded from Fig. 20.

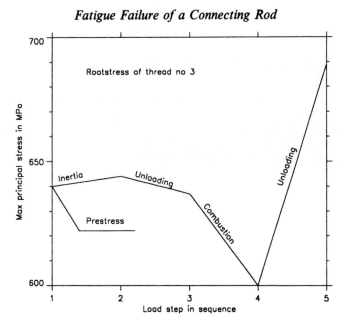

Fig. 19. Negative effect of insufficient load cycling.

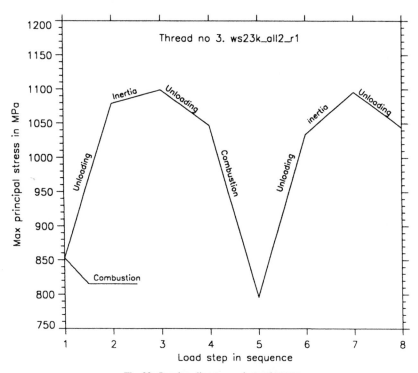

Fig. 20. Load cycling to reach steady state.

7. CONCLUSIONS

The FE analyses and the results of fatigue testing led to the following solutions to the fatigue problem:

(a) The screw thread profile was modified to incorporate a root radius specified equal to 0.3 mm.
(b) The material of the connecting rod was changed to 34CrNiMo6 TQ +T EN 100f3-1 because of its increased fatigue strength.
(c) The screws were phosphated in order to reduce the risk of fretting.

NOTCH STRESS-STRAIN ANALYSIS IN MULTIAXIAL FATIGUE

Meera N.K. Singh[1], Andrzej Buczynski[2], and Grzegorz Glinka[1]

[1]Department of Mechanical Engineering
The University of Waterloo
Waterloo, Ontario, Canada, N2L 3G1
[2]Warsaw University of Technology
Warsaw, Poland

Abstract. An analytical method for calculating notch tip stresses and strains in elastic-plastic isotropic bodies subjected to multiaxial non-proportional loading sequences is presented. The key elements of the two proposed models for monotonic loading are generalized relationships between fictitious elastic and elastic-plastic strain energy densities, and the material constitutive relations. Each method consists of a set of seven linear algebraic relations that can easily be solved for elastic-plastic strain and stress increments knowing the hypothetical notch tip elastic stress history and the material stress-strain curve. Results of the validation show that the proposed methods compare well with finite element data and that each solution set forms the limits of a band within which actual notch tip strains fall. The use of the proposed models in conjunction with cyclic plasticity models is discussed for application to notched bodies subjected to cyclic loading sequences.

Keywords. Notch Analysis, Neuber's Rule, Strain Energy Density Method, Multiaxial Fatigue, Nonproportional Loading, Cyclic Plasticity, Mroz Model.

INTRODUCTION

Fatigue durability and strength analysis of machine components and structures subjected to multiaxial cyclic loads requires the determination of elastic-plastic strains and the fatigue damage accumulated at the point of the highest stress concentration. For this purpose, models that efficiently simulate notch tip stress-strain histories due to externally applied cyclic loads are required. Once developed, the resulting histories can be used to determine the amount of fatigue damage at the notch, providing that an appropriate damage parameter is available.

The most frequently used methods for calculating the notch tip stress-strain field due to cyclic loads are Neuber's rule [1] which has been extended to fatigue problems by Topper et. al [2] and the equivalent strain energy density (ESED) method [3] [4]. An extension of Neuber's rule for multiaxial stress states has been proposed by Hoffman and Seeger [5] and recently by Barkey and Socie [6]. A more general extension of Neuber's rule and the ESED method for multiaxial loading has been proposed by Moftakhar et. al. [7] [8]. The method proposed in [7] and [8] is based on strain energy density considerations and is governed by the assumption that the multiaxial loads are applied in a proportional manner. A similar formulation, appropriate for notched bodies subjected to multiaxial non-proportional loads applied monotonically, can be found in references [9] and [10].

In this paper, the incremental models given in [9] and [10] for monotonic loading are discussed and the method by which they can be applied to notched bodies subjected to non-proportional loading where both loading and unloading reversals occur is presented.

STRAIN ENERGY DENSITY MODELS FOR MULTIAXIAL NON-PROPORTIONAL LOADING

Basic Assumptions and Relations

If the body dimensions and external loads are such that the body is in the state of plane stress, the stress state at the notch tip is uniaxial (Fig. 1a). Four independent relations are required to define the unknown notch tip stress and three strain components. Similarly if the notched body is in the plane strain state (Fig. 1b), four relations are needed to determine the four unknown notch tip strain and stress components. Three independent relations can be defined by the material constitutive equations and thus only one additional relation is required.

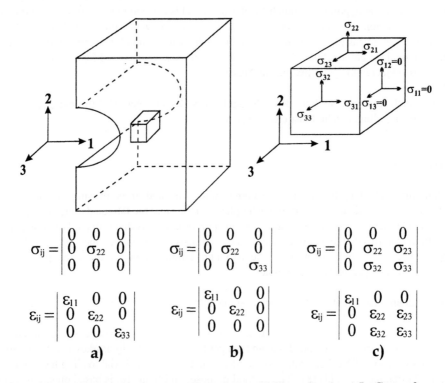

$$\sigma_{ij} = \begin{vmatrix} 0 & 0 & 0 \\ 0 & \sigma_{22} & 0 \\ 0 & 0 & 0 \end{vmatrix} \qquad \sigma_{ij} = \begin{vmatrix} 0 & 0 & 0 \\ 0 & \sigma_{22} & 0 \\ 0 & 0 & \sigma_{33} \end{vmatrix} \qquad \sigma_{ij} = \begin{vmatrix} 0 & 0 & 0 \\ 0 & \sigma_{22} & \sigma_{23} \\ 0 & \sigma_{32} & \sigma_{33} \end{vmatrix}$$

$$\varepsilon_{ij} = \begin{vmatrix} \varepsilon_{11} & 0 & 0 \\ 0 & \varepsilon_{22} & 0 \\ 0 & 0 & \varepsilon_{33} \end{vmatrix} \qquad \varepsilon_{ij} = \begin{vmatrix} \varepsilon_{11} & 0 & 0 \\ 0 & \varepsilon_{22} & 0 \\ 0 & 0 & 0 \end{vmatrix} \qquad \varepsilon_{ij} = \begin{vmatrix} \varepsilon_{11} & 0 & 0 \\ 0 & \varepsilon_{22} & \varepsilon_{23} \\ 0 & \varepsilon_{32} & \varepsilon_{33} \end{vmatrix}$$

a)   b)   c)

Fig. 1: State of Stress at the Notch Tip a) Plane Stress, b) Plane Strain, c) In General

The additional energy relation required for the plane stress or plane strain state is generally defined by either Neuber's rule [1] or the ESED [3] relation given respectively as:

$$\sigma_{22}^{e}\varepsilon_{22}^{e} = \sigma_{22}^{N}\varepsilon_{22}^{N}, \tag{1}$$

$$\frac{1}{2}\sigma_{22}^{e}\varepsilon_{22}^{e} = \int_{0}^{\varepsilon_{22}^{\varepsilon}}\sigma_{22}^{E}\,d\varepsilon_{22}^{E}, \tag{2}$$

where the superscripts N and E refer to elastic-plastic values estimated by Neuber's rule and the ESED method respectively. Both models relate the fictitious "linear elastic" strains and stresses at the notch tip $(\sigma_{ij}^{e},\varepsilon_{ij}^{e})$ to the elastic-plastic strains and stresses $(\sigma_{ij},\varepsilon_{ij})$, as shown in Fig. 2.

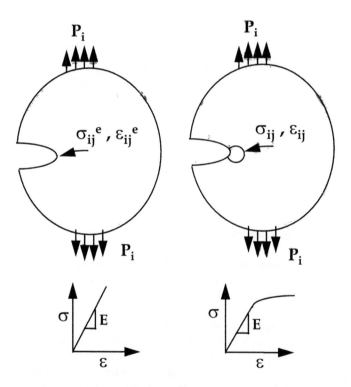

Fig. 2: Geometrically Identical Elastic and Elastic-Plastic Bodies Subjected to the Same Boundary Conditions.

In the case of a general multiaxial loading applied to a notched body, the stress state near the notch tip is triaxial, and at the notch tip, biaxial due to the free surface (Fig. 1c). Since equilibrium of the element at the notch tip has to be observed, then components $\sigma_{23} = \sigma_{32}$ and $\varepsilon_{23} = \varepsilon_{32}$, and there are seven unknowns at the notch tip: three stress and four strain components. Therefore, a set of seven independent equations is required to completely define the stress-strain state at the notch tip. The material constitutive relations provide four equations, and thus three additional relations are required.

M. N. K. Singh et al.

In the case of plastic deformation, the final stress-strain state is dependent on the loading path. As such, relations that define the local stress-strain state in a notched body subjected to a multiaxial loading system must be developed in an incremental form. In other words, three stress increments and four strain increments have to be found for each load increment. The constitutive incremental equations form four independent equations. The remaining three equations necessary for a complete formulation of the notch tip problem can be determined by using the strain energy criteria discussed in references [9] and [10], and presented below.

Material Constitutive Model

The most frequently used material constitutive model used when external loads are applied such that the stresses are increased in a body in a nonproportional manner is the Prandtl-Reuss relation. For an isotropic body, the Prandtl-Reuss relations can be expressed as:

$$\Delta \varepsilon_{ij} = \frac{1+\nu}{E} \Delta \sigma_{ij} - \frac{\nu}{E} \Delta \sigma_{kk} \delta_{ij} + \frac{3}{2} \frac{\Delta \varepsilon_{eq}^{p}}{\sigma_{eq}} S_{ij}, \tag{3}$$

$$\sigma_{eq} = \sqrt{\frac{3}{2} S_{ij} S_{ij}} \;\; ; \;\; S_{ij} = \sigma_{ij} - \frac{1}{3} \sigma_{kk} \delta_{ij},$$

$$\Delta \varepsilon_{eq}^{P} = \frac{df(\sigma_{eq})}{d\sigma_{eq}} \Delta \sigma_{eq},$$

for any stress strain relation such as: $\varepsilon_{eq}^{p} = f(\sigma_{eq})$. Here, i, j = 1, 2, 3, and the summation convention applies.

Generalized Strain Energy Relations

Simplified models [1]-[11] generally make use of fictitious elastic notch tip stress-strain data to predict the actual elastic-plastic notch tip behaviour. Two models are proposed here to address notched bodies subjected to multiaxial non-proportional loads.

Equivalent Strain Energy Density (ESED) Relations. If an increment of load is applied to a body, there will be a corresponding incremental increase in the actual strain energy in the body. Consider the notch tip element shown in Figure 1c. Assume that (i) the remote stresses are elastic, (ii) the notch tip strains are elastic-plastic, and (iii) the notch tip behaviour is largely controlled by the surrounding elastic field. Under these assumptions, it is proposed that for a given increment of external load, the corresponding increment in the strain energy density at the notch tip in an elastic-plastic body can be approximated by that which would be obtained if the body was to hypothetically remain elastic throughout the loading history. This hypothesis can be expressed as:

$$\Delta W^{e} = \Delta W^{E}$$

or \tag{4}

$$\sigma_{ij}^{e} \Delta \varepsilon_{ij}^{e} = \sigma_{ij}^{E} \Delta \varepsilon_{ij}^{E}.$$

Equation (4) is called the incremental equivalent strain energy density method since it reduces to the original form of the ESED method (eqn. 2) for uniaxial notch tip stress states. Furthermore, it represents a statement of equality between the increment of notch tip strain energy density obtained from a linear elastic solution and that obtained from an elastic-plastic analysis. A graphical representation of the incremental ESED method is shown in Figure 3a.

The four constitutive relations used in conjunction with the generalized ESED equation will only be sufficient to formulate a set of five equations. Therefore, two more independent equations are required to completely define the notch tip stress-strain increments for a given increment in the applied load. Therefore, the following hypothesis is made that states that the equality of energies applies also to all corresponding stress and strain components:

$$\sigma_{\alpha\beta}^{e} \Delta \varepsilon_{\alpha\beta}^{e} = \sigma_{\alpha\beta}^{E} \Delta \varepsilon_{\alpha\beta}^{E}. \tag{5}$$

Note that in eqn. (5), the indices α, $\beta = 1, 2, 3$, and summation is not implied. The three relations implied by eqn. (5) and the four by the constitutive equations are sufficient to determine the three unknown stress and four unknown strain increments. A similar formulation of the generalized ESED method has also been proposed by Chu and Conle [11].

Generalized Neuber's Rule. If a load increment is applied to a body, there will be a corresponding incremental increase in the total strain energy in the body. The total strain energy refers to the sum of the strain energy density (as described above) and the complementary strain energy density. Consider the notch tip element shown in Figure 1c. Assume that the nominal remote stresses are elastic, the notch tip strains are elastic-plastic, and the notch tip behaviour is largely controlled by the surrounding elastic field. Under these assumptions, it is proposed that for a given increment in an external load, the corresponding increment in the total strain energy density at the notch tip in an elastic-plastic body can be approximated by that which would be obtained if the body was to hypothetically remain elastic throughout the loading history. Mathematically, this can be written as:

$$\Delta \Omega^{e} = \Delta \Omega^{N}$$

$$\text{or} \tag{6}$$

$$\sigma_{ij}^{e} \Delta \varepsilon_{ij}^{e} + \varepsilon_{ij}^{e} \Delta \sigma_{ij}^{e} = \sigma_{ij}^{N} \Delta \varepsilon_{ij}^{N} + \varepsilon_{ij}^{N} \Delta \sigma_{ij}^{N}.$$

Equation (6) is called the incremental Neuber's relation since it reduces to Neuber's rule in its original form (eqn.1) for uniaxial notch tip stress states. Furthermore, it represents a statement of equality between the increment in the notch tip total strain energy density obtained from a linear elastic solution and that obtained from an elastic-plastic analysis. A graphical representation of the incremental Neuber's rule is shown in Fig. 3b.

The four constitutive relations used in conjunction with the generalized Neuber's rule (eqn. 6) will only be sufficient to describe five of the seven unknown notch tip stress-strain increments. Therefore, two more independent equations are required to completely define the notch tip stress-strain increments for a given increment in the applied load. Again, it is proposed that the contribution of each elastic-plastic stress-strain component to the total strain energy density at the notch tip is the same as the contribution of analogous stress-strain component to the total strain energy density at the notch tip assuming that the body was to remain elastic during the loading history. This proposal can be expressed as:

$$\sigma^e_{\alpha\beta} \Delta \varepsilon^e_{\alpha\beta} + \varepsilon^e_{\alpha\beta} \Delta \sigma^e_{\alpha\beta} = \sigma^N_{\alpha\beta} \Delta \varepsilon^N_{\alpha\beta} + \varepsilon^N_{\alpha\beta} \Delta \sigma^N_{\alpha\beta}. \tag{7}$$

The three relations implied by eqn.(7) and four constitutive equations form a set of seven independent equations sufficient for determining the unknown increments $\Delta\sigma_{ij}{}^N$ and $\Delta\varepsilon_{ij}{}^N$.

In order to determine the notch tip elastic-plastic strains and stresses at the end of the loading history, they must first be evaluated for each increment in the applied load. Initially, the first reference state is taken as the point at which yielding occurs at the notch tip since it can be found from an elastic analysis of the body. For each increment in external load, the increments in the elastic-plastic notch tip strains and stresses are computed from either eqns. (3) and (5), or eqns. (3) and (7) with the knowledge of the hypothetical elastic notch tip stress-strain history and the material stress-strain curve. The stress and strain states at the end of given load increment is then computed using:

$$\sigma^n_{ij} = \sigma^o_{ij} + \sum_{k=1}^{n-1} \Delta\sigma_{ij} + \Delta\sigma^n_{ij}, \tag{8}$$

$$\varepsilon^n_{ij} = \varepsilon^o_{ij} + \sum_{k=1}^{n-1} \Delta\varepsilon_{ij} + \Delta\varepsilon^n_{ij}, \tag{9}$$

where n denotes the number of the load increment.

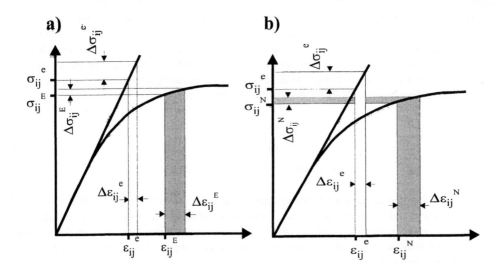

Fig. 3: Graphical Representation of the a) Incremental ESED Method and b) Incremental Neuber's Rule

INCREMENTAL CYCLIC PLASTICITY MODEL

The set of equations defined by the incremental Neuber's rule and ESED method can be solved only if the relation between the equivalent plastic strain increment $\Delta\varepsilon_{eq}{}^P$ and the equivalent stress $\Delta\sigma_{eq}$ is known during the application of given load increment. However, it is known that the current $\Delta\varepsilon_{eq}{}^P$ -

$\Delta\sigma_{eq}$ relation depends on the previous load path and therefore the use of the incremental Neuber's rule or the ESED method has to be associated with a plasticity model dealing with path dependent material constitutive behaviour. Several models [12]-[15] are available in the literature of which the model proposed by Mroz [12] and recently improved by Garud [13] are the most popular.

Mroz [12] has proposed that the uniaxial stress-strain material curve can be represented by a set of plasticity surfaces in three-dimensional stress space (Fig. 4). In the case of a two-dimensional stress state, as it is at the notch tip, the plasticity surfaces reduce to ellipses on the plane of principal stresses described by:

$$\sigma_{eq} = \sqrt{\sigma_1^2 - \sigma_1\sigma_2 + \sigma_2^2}. \tag{10}$$

The load path dependent memory effects are modelled by prescribing a translation rule for the ellipses moving with respect to each other over distances given by the stress increments. It is also assumed that the ellipses move inside each other and they do not intersect. If two ellipses come in contact with one other they move together as one rigid body.

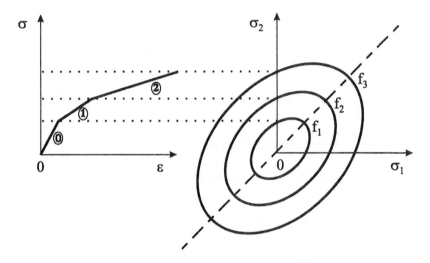

Fig. 4: Piecewise Linearization of the Material σ-ε Curve and the Corresponding Elliptical Plasticity Surfaces

The translation rule proposed by Garud [13] avoids the intersection of the ellipses that could occur in some cases in the original Mroz [12] model. The Garud translation rule is illustrated in Fig. 5 and can be described by a model having, for simplicity, only two plastic surfaces (two ellipses).

M. N. K. Singh et al.

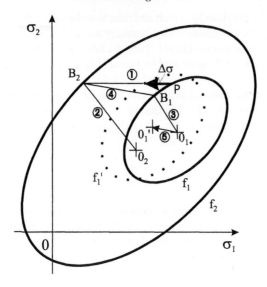

Fig. 5: Geometrical Illustration of the Mroz-Garud Incremental Plasticity Model

In order to model material response due to the stress increment $\Delta\sigma$, the following steps are to be followed.

1. Extend the line of action of the stress increment $\Delta\sigma$ to intersect the external non-active plastic surface f_2 at point B_2.

2. Connect point B_2 and the centre O_2 of the plastic surface (ellipse) f_2.

3. Draw through the centre O_1 of the smaller active ellipse a line parallel to the line O_2B_2 to find point B_1 on the plastic surface f_1.

4. Connect the conjugate points B_1 and B_2 by line B_1B_2.

5. Translate ellipse f_1 from point O_1 to O_1' in such a way that vector O_1O_1' is parallel to the line B_1B_2. The translation is completed when the end of the vector $\Delta\sigma$ is found on the translated ellipse f_1'.

The translation rule described above assures that the two ellipses are tangential with the common point B_1B_2 without intersecting each other. Two or more tangent ellipses translate together as rigid bodies and the largest moving ellipse (Fig. 4) indicates which constitutive relation (which linear piece) should be used for a given stress increment.

In most publications, the plasticity models are described as algorithms for calculating strain increments due to given stress increments or vice versa. However, in the case of the incremental Neuber or ESED method, both the strain and stress increments are determined. Therefore, the plasticity model is needed only for indication which ellipse is going to be active during the next notch tip stress increment due to the fictitious elastic stress increments $\Delta\sigma_{ij}^e$. This can be determined based on the actual configuration of all plastic surfaces and the increment or decrement of the equivalent elastic stress $\Delta\sigma_{eq}°$, which can be determined as a differential from eqn. (3). Therefore, the use of the multi-surface plasticity model can be reduced to an appropriate translation of the plastic surfaces after each load increment.

The Garud model [13] was chosen here as an illustration but any other plasticity model can be associated with the incremental Neuber and ESED method.

VALIDATION OF PROPOSED MODELS

The accuracy of the proposed incremental ESED method and incremental Neuber's rule are assessed by comparing notch tip stress-strain histories obtained using the models to those obtained using elastic-plastic finite element data [16]. The FEM data was generated using the incremental flow rule and the equivalent stress throughout the body was increasing under the given loads. The geometry of the chosen model is that of the circumferentially notched bar shown in Fig. 6a. Loads applied to the bar were chosen to be tension and torsion, applied along various non-proportional monotonic loading paths. Here, results are presented only for the path shown in Fig. 6b. The applied torsional $\tau_{n.s.}$ and tensile $\sigma_{n.s.}$ stress are determined as follows for the net cross section and they represent the applied loads:

$$\sigma_{n.s.} = \frac{P}{\pi(R - t)^2} \; ; \; \tau_{n.s.} = \frac{2T}{\pi(R - t)^3}. \tag{11}$$

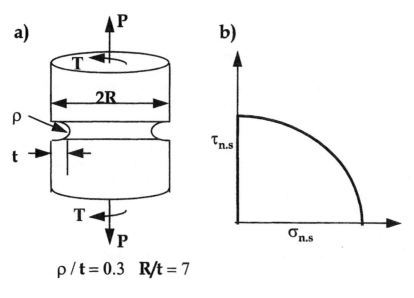

$$\rho / t = 0.3 \quad R/t = 7$$

Fig. 6: Numerical Assessment a) Geometry of Model, b) Loading Path

The material of the cylinder was chosen to be the SAE 1045 steel having the cyclic stress-strain curve given by Ramberg-Osgood relation:

$$\varepsilon = \frac{\sigma}{E} + \left(\frac{\sigma}{K'}\right)^{\frac{1}{n'}},$$

(12)

with parameters E = 202 GPa, υ = 0.3, S_y = 202 MPa, n' = 0.208, and K' = 1258 MPa.

The elastic notch tip stress histories were determined by a simple elastic finite element analysis. In doing so, it was found that for any given nominal tensile stress $\sigma_{n.s.}$, or any given nominal torsional stress, $\tau_{n.s.}$, the resulting elastic notch tip stress components were: $\sigma_{22}^e = 3.31\sigma_{n.s.}$, $\sigma_{33}^e = 0.94\sigma_{n.s.}$, and $\sigma_{23}^e = 1.94\tau_{n.s.}$.

The maximum applied torsional load (Fig. 6) was chosen to be 50% higher than would be required to cause yielding at the notch tip ($\tau_{n.s.}^f$ = 90 MPa). The torsional load was then reduced and the tensile load was increased in a manner such that the equivalent elastic notch tip stress remained constant and equal to $\sqrt{3}(1.94)\tau_{n.s.}^f$ for the remainder of the loading history.

The notch tip elastic stress histories induced by the load path shown in Fig. 6 were used in eqns. (3) and (5) (ESED) and (3) and (7) (Neuber's) to calculate the notch tip elastic-plastic strains and stresses. A short computer program was written for this purpose. The largest stress σ_{22}, σ_{23} (Fig 7a,b) and strain ε_{22} and ε_{23} (Fig 8a,b) components determined using the incremental solution sets and the elastic-plastic finite element model, are plotted together against the normalized nominal equivalent stress defined as:

$$\frac{S_{eq}}{S_y} = \frac{\sqrt{\sigma_{n.s.}^2 + 3\tau_{n.s}^2}}{S_y}.$$

(13)

It is noted that, both models and the finite element results are identical in the elastic range. This is expected since in the elastic range, all models converge to the elastic solution. Just after the onset of plastic yielding at the notch tip, the strain results predicted using each proposed model and the finite element results begin to diverge. The divergence becomes more pronounced with the increase of the nominal equivalent stress S_{eq}.

In the notch tip stress plots (Fig. 7), it can be immediately noted that both simplified models predict the general trend of the finite element stress history. In both plots, it is further noted that the incremental Neuber's solution set provides a better prediction of the finite element results than does the generalized ESED solution set. However, the largest error occurs at the end of the loading paths, where the assumption of localized plasticity is violated.

The notch tip strain plots (Fig. 8) show that the incremental solution sets predict the general trend in the finite element strain histories. It is noted that the generalized Neuber's model predicts higher strains than does the generalized ESED method and the models bound the finite element results. It can be then said that in the case of strain components, Neuber's rule predicts an upper bound, and the generalized ESED method, a lower bound approximation to the actual notch tip strains. This conclusion is consistent with numerous other numerical validation models conducted.

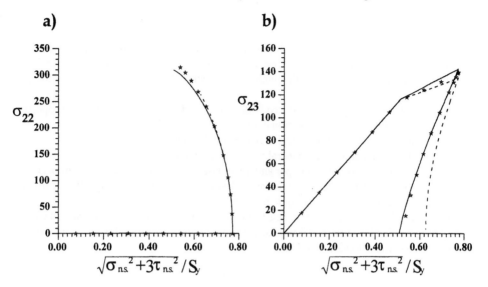

Fig. 7: Notch Tip Stress Histories (MPa) a) Parallel, b) Transverse to Bar Axis
(—— **Incremental Neuber's Rule, ★★★ FEM Results, ········ Incremental ESED Method**)

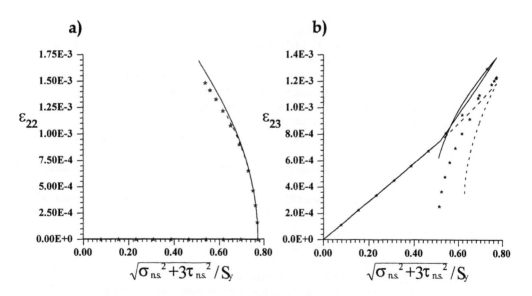

Fig. 8: Notch Tip Strain Histories a) Parallel, b) Transverse to Bar Axis
(—— **Incremental Neuber's Rule, ★★★ FEM Results, ········ Incremental ESED Method**)

CONCLUSIONS

The proposed solution methodologies enable the complete elastic-plastic notch tip stress-strain state in a body subjected to a non-proportional loading sequence to be estimated based on the material's uniaxial stress-strain curve, and the elastic notch tip stress-strain history. The benefits of the proposed models are

- They can be applied to notched bodies subjected to uniaxial, multiaxial proportional, and multiaxial non-proportional loading paths.
- The equations are linear, and they are easy for numerical analysis.
- The models predict the general trend in the actual stress-strain results.
- The models provide an upper and lower bound solution on the actual strains and the resulting band is relatively narrow. The efficiency of the strain results is particularly important for use with fatigue life prediction models.

Finally, it is shown that the simplified models presented can be used with relative ease in conjunction with any chosen cyclic plasticity model to predict the notch tip stress-strain history in a body subjected to a multiaxial non-proportional cyclic load.

REFERENCES

1. Neuber, H. (1961). *ASME Journal of Applied Mechanics*, **Vol.26 No.4**, pp. 544-550.
2. Topper, T.H., Wetzel, R.M. and Morrow, J. (1969). *Journal of Materials*, **Vol.5 No.1**, pp. 200-290.
3. Molski, K., and Glinka, G. (1981). *Material Science and Engineering*, **Vol.50**, pp. 93-100.
4. Glinka, G. (1985). *Engineering Fracture Mechanics*, **Vol.22 No.5**, pp. 839-854.
5. Hoffmann, M., Amstutz, H., and Seeger, T. (1991). In: *Fatigue Under Biaxial and Multiaxial Loading ESIS10*, K. Kussmaul, D. McDiarmid, D. Socie (Eds). Mechanical Engineering Publications, London, pp. 357-376.
6. Barkey, M.E., Socie, D.F., and Hsai, K.J. (1993). In: *Theoretical and Applied Mechanics Report No. 709*, University of Illinois at Urbana, Urbana, Illinois.
7. Moftakhar, A., Buczynski, A., and Glinka, G. (1993). In: *Fatigue '93*, J.P. Bailon, and J.I. Dickson (Eds), EMAS, **Vol.1**, pp. 441-452.
8. Moftakhar, A., Buczynski, A., and Glinka, G.(1995). *International Journal of Fracture*, **Vol. 70**, pp. 357-373.
9. Singh, M.N.K., Glinka, G., and Dubey, R.N. (1996). *International Journal of Fracture*, **Vol.76**, pp. 39-60.
10. Singh, M.N.K., Moftakhar, A., and Glinka., G. (1995). In: *Computational Mechanics '95 Theory and Applications*, A.N. Atluri, G. Yagawa, T.A. Cruse (Eds) ,Springer-Verlag, Heidelberg, Germany, pp. 1316-1323.
11. Chu, C.C., and Conle, F.A. (1994). In: *The Proceedings of the Fourth International Conference on Biaxial/Multiaxial Fatigue*, Paris, France.
12. Mroz, Z. (1967). *Journal of the Mechanics and Physics of Solids*, **Vol.15**, pp. 163-175.
13. Garud, Y.S.(1981). *ASME Journal of Engineering Materials and Technology*, **Vol.103**, pp. 118-125.
14. Chu, C.C. (1984). *Journal of the Mechanics and Physics of Solids*, **Vol.22 No.3**, pp. 197-212.
15. Jiang, Y., and Kurath, P.(1996). *International Journal of Plasticity*, **Vol.12 No.3**, pp. 387-415.
16. Hibbit, Karlsson & Sorensen, Inc.(1993). *ABAQUS Theory Manual*, Version 5.3.

FRACTURE MECHANICAL FATIGUE ANALYSIS
OF RAILWAY WHEELS WITH ROLLING DEFECTS

K.-O. Edel and G. Boudnitski
Fachhochschule Brandenburg
D-14770 Brandenburg an der Havel, Germany

ABSTRACT

Reusable solid railway wheels manufactured of the old wheel steel BV 1 (nearly equivalent to the solid or monobloc wheel steel R1 according to the new nomenclature) show in some cases rolling defects on the surface of the wheel disc. The fatigue behaviour of such defects is analysed by means of the linear-elastic fracture mechanics and Monte Carlo simulation and assessed to be able to derive the allowable defect size for the non-destructive testing.

KEYWORDS

fracture mechanics, fatigue, threshold ranges of the stress intensity factor, railway wheel, rolling defects, Monte Carlo simulation, safety factors, allowable defect size.

NOMENCLATURE

r radial coordinate
r_{norm} normal distributed random number with the mean 0 and the standard deviation 1
$s_{\sigma b}$ standard deviation of the braking stresses in radial direction
$s_{\sigma r}$ standard deviation of the residual stresses in radial direction
t_{rim} thickness of the rim of the solid wheel
α inclination angle between the crack area and the surface of the wheel disc
σ_{brake} braking stress in radial direction
σ_{res} residual stress in radial direction

INTRODUCTION

The rolling stock of the railways usually has a limited service life of some decades. At the end of their life the waggons will be scrapped. But not all parts of the railway structures will be unusable at this time. In many cases the wheels show only a very small wear on the tread so that the wheel sets are reusable.

125

At the reconditioning of the reusable wheel sets the wheels will be tested non destructively. Circumferentially and radially directed surface defects on the wheel disc were found in some cases. According to metallographic micro slices these defects are rolling defects with a finite or infinitely small tip radius.

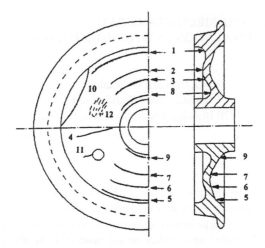

Fig. 1. The different kinds of rolling defects on the surface of the wheel disc.

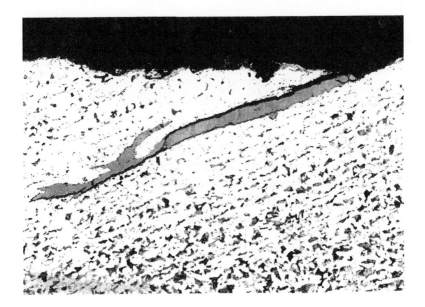

Fig. 2. Micro slice transverse to the defect [1].

For the following investigations the rolling defects are assumed as cracklike defects with an infinitely small crack tip radius. Under this realistic assumption it is possible to use the linear elastic fracture mechanics to analyse and to assess the behaviour of these defects.

GEOMETRY OF DEFECTS

The tangentially directed defects which are of special interest show a length L of up to 600 *mm*. The inclined crack size a under the surface of the wheel disc has a size of up to about 10 *mm*. The correlation between the defect length L on the surface and the crack size a under the surface is given in Fig. 3.

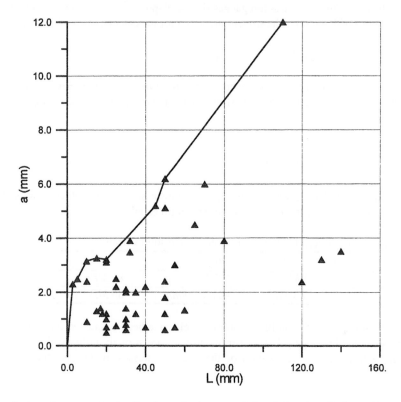

Fig. 3. Correlations between the inclined crack size a and the defect length L on the surface of the wheel disc.

The inclination of the crack area against the surface of the wheel disc lies between 0° and 90° with a mean value of about 25°. The probability of the inclination angle is approximated by the equation

$$P(\alpha) = 1 - \exp\left[-\left(\frac{\alpha - 8.46°}{18.01°}\right)^{1.11}\right]. \tag{1}$$

STRESSES AND STRESS INTENSITY FACTORS

Stresses in the Wheel Disc [2]

The wheel force components which act on the rim of the wheel depend on the track and rail geometry. The vertical wheel force component F_Q acts on the tread. It has a size of up to 170 *kN* for a load capacity of the wheel set of 20 tons. The horizontal wheel force component F_Y acts either on the flange ($F_Y > 0$) or on the inner side of the rim ($F_Y < 0$). It is scattered in the range of -23 up to 55 kN for the same load capacity.

The stresses in the wheel disc are measured under static conditions for predetermined wheel force components. The positioning of the strain gauges -in the $0°$-position - were between the wheel set axle and the contact point with the rail. The stresses are given for the calculations by the equations

$$\sigma_{r,0°} = c_Q \cdot F_Q + c_Y \cdot F_Y \qquad \text{and} \qquad \sigma_{r,180°} = -c_Y \cdot F_Y. \tag{2}$$

Residual stresses in the disc of some solid wheels have been measured by means of strain gauges. Depending on the radius r of the position of the strain gauge, the residual stresses are approximated by the equation

$$\sigma_{res} = \sigma_{res}(r = 0) + \frac{d\sigma_{res}}{dr} \cdot r + s_{\sigma r} \cdot r_{norm} \tag{3}$$

with the regression coefficients given in Table 1. The radius r is limited to values between 171 mm and 317 mm corresponding to the extreme positions of the strain gauges.

Table 1. Coefficients for describing the residual stresses in solid wheels.

surface	$\sigma_{res}(r = 0)$	$d\sigma_{res}/dr$	$s_{\sigma r}$
outher	-218.45 MPa	0.7375 MPa/mm	80 MPa
inner	+249.59 MPa	-1.2484 MPa/mm	80 MPa

Thermal stresses in the disc of tread braked solid wheels have been calculated for a differently shaped wheel under usual stop brakings [3]. The thermal stresses are strongly influenced by the thickness of the rim of the wheel. For the calculation of the braking stresses in the disc the following rough approximation is used

$$\sigma_{brake} = \sigma_{brake}(t_{rim} = 0) + \frac{d\sigma_{brake}}{dt_{rim}} \cdot t_{rim} + s_{\sigma b} \cdot r_{norm.} \tag{4}$$

with the coefficients given in Table 2 and the thicknesses of the rim between 16 and 93 mm.

Table 2. Coefficients for describing the braking stresses in radial direction.

$\sigma_{brake}(t_{rim} = 0)$	$d\sigma_{brake}/dt_{rim}$	$s_{\sigma b}$
229.74 MPa	-1.2338 MPa/mm	22.4 MPa

Stress Intensity Factors

Solutions for the plane problem of straight inclined edge cracks loaded by constant tension (membrane stresses) and bending stresses are given in the handbook published by MURAKAMI [4]. The most complete solution of the mixed mode problem is given by BOWIE [5] for membrane stresses. The solutions to the equivalent bending problem show a reduction of the calibration factors of up to 50 per cent for the same nominal stresses.

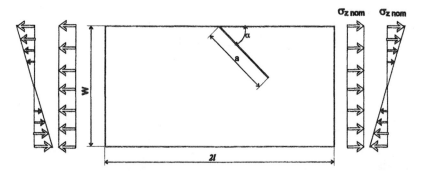

Fig. 4. Fracture mechanical idealization of the rolling defects as straight inclined edge cracks under the action of membrane and bending stresses.

To calculate the stress intensity factors the thickness W of the wheel disc must be known. With r und W in *mm* the thickness is approximated by

$$W = 38.2 - 0.447 \cdot r. \tag{5}$$

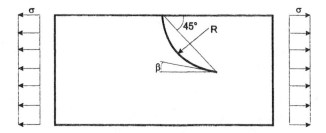

Fig. 5 Fracture mechanical idealization of the rolling defects to investigate the influence of the defect curvature.

Test calculations using the crack model given in Fig. 5 show for crack tip angles $\beta < 45°$ a reduction of the equivalent stress intensity factor compared with the straight crack and for crack tip angles $\beta > 45°$ an increase of up to about 3 per cent over that value of straight cracks. For crack tip directions which tends towards parallel to the surface the model of the straight inclined crack is conservative. For crack tip directions which tends towards transverse to the surface the model of the straight inclined crack is a little bit unsafe.

THRESHOLD VALUES ΔK_{th} OF WHEEL STEEL R1

Only a few published threshold ranges ΔK_{th} of the stress intensity factor for solid wheel steel are known [6]. Most investigations are related to the steels R7 and R9 which are used for wheels with quenched rims [7]. Because of the different carbon content and the heat treatment of the steels investigated before and the old not heat treated steel R1 it is necessary to determine the threshold values ΔK_{th} of the steel R1. The results ΔK_{th} of about 60 tests are given in Fig. 6 against the stress ratio R.

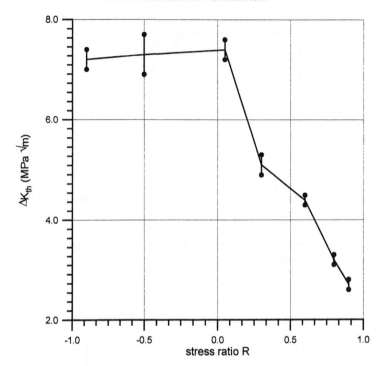

Fig. 6. Mean value and standard deviation of the threshold range ΔK_{th} of the stress intensity factor of the solid wheel steel BV 1 (or R1).

Table 3. Chemical composition of wheel steels [8, 9].

steel	chemical composition in weight per cent				
	C	Si	Mn	P	S
BV1 [8]	---	0.50	1.20	0.05	0.05
R1 [9]	0.48	0.50	0.90	0.035	0.035
R7 [9]	0.52	0.40	0.80	0.035	0.035
R9 [9]	0.60	0.40	0.80	0.035	0.035

MONTE CARLO SIMULATION OF FATIGUE

To compare the material loading with the fracture mechanical properties the stress intensity factors of the mode I and II can be summed up to an equivalent value of the stress intensity factors. Fracture tests [10] and crack growth tests show a good result using the equation

$$K_{I,V} = \frac{1}{2} \cdot \left(K_I + \sqrt{K_I^2 + 6 \cdot K_{II}^2} \right). \tag{6}$$

It can be assumed that the above mentioned equation is also suited for the case of not growing cracks. The fatigue strength condition is then

$$\Delta K_{I,V} = \Delta K_{th}. \tag{7}$$

The fracture mechanical calculations of the crack size a at which the cyclic growth begins are performed with random values of the essential properties, parameters, and dimensions. The calculations are repeated 5000 times for the same assumptions to get the probability distribution. As it is unknown how large the membrane stress part and the bending stress part are in the disc the simulation is performed for two extreme assumptions: on the one hand for pure membrane stresses and on the other hand for pure bending stresses (with the same stress values but a calibration factor in the equations for the stress intensity factors which is reduzed to 50 per cent compared with that of the membrane stresses).

The Monte Carlo simulation gives three different results:
- For stresses in the pressure range it can be assumed that the cracklike defects cannot grow. (The effect of the small K_{II} will be neglected.) This case won't diminish the safety of the wheels under practical conditions.
- For low stresses in the tension range the threshold crack size can exceed the thickness of the wheel disc. This case likewise won't diminish the safety of the wheels under practical conditions.
- For relatively high stresses in the tension range there exists a threshold crack size between non-growing and growing.

All possible cases are considered appropriate to their probability to get the survival probability of the wheels (Fig. 7). To get a representative value of the non-growing defect size that defect size is selected

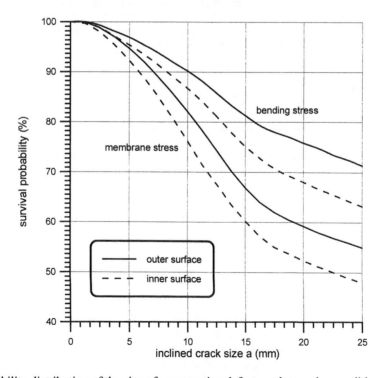

Fig. 7. Probability distribution of the size of non growing defects under service conditions.

from the simulated probability distribution which has a survival probability of 90 per cent. To be independent of the assumptions related to the stress distribution the results of the membrane stress and bending stress calculations are averaged. But this assumption ist conservative. More realistic is the assumption of a state of bending stresses. Consequently, on the inner surface of the wheel disc follows

$$a(P_D = 90\%) = \begin{cases} 7{,}0 \; mm & \text{for the mean of membrane and bending stresses} \\ 8{,}2 \; mm & \text{for bending stresses} \end{cases} . \tag{8}$$

ASSESSMENT OF THE SIMULATION RESULTS

Safety factors to guarantee the durability of cyclic loaded structures which are usually related to the stresses are given in Table 4. For the use in fracture mechanical calculations and assessments, the safety factors must be modified according to the relation

$$S_a = S_\sigma^2 \tag{9}$$

to get safety factors which are related to the crack size.

Table 4. Safety factors S_σ for cyclic loaded structures manufactured of rolled steel [11].

consequences of failure	large	small
large welded structures with not considered residual stresses:		
not periodical NDT	1.9	1.6
periodical NDT	1.7	1.5
other cases of structures:		
not periodical NDT	1.5	1.3
periodical NDT	1.35	1.2

The necessary safety factor to assess the rolling defects has a size of $1{,}35^2 = 1{,}82$. The use of the before mentioned safety factor to the representative minimum value of the not growing defects leads to the allowable crack size

$$a_{all.} = \frac{a(P_D = 90\%)}{S_a} = \begin{cases} 3.8 \; mm & \text{for the mean of membrane and bending stresses} \\ 4.5 \; mm & \text{for bending stresses.} \end{cases} \tag{11}$$

CONCLUSIONS

The allowable value of the crack size a, which was calculated by means of the LEFM, is not very suitable for the non-destructive testing of the surface of the wheel disc. The length L of the defect on the surface is essentially better suited for the NDT than the inclined crack size a. Starting from the above derived allowable crack size a the allowable value of the crack length L is derived by means of the border line in Fig. 3 to about 40 mm but not smaller than 30 mm. The assessment of rolling defects by means of this allowable defect length does not requires the determination of the defect size below the surface of the wheel disc. The explained method to derive the allowable defect size can likewise be applied to other cyclic loaded structures.

ACKNOWLEDGEMENTS

The investigations presented were initiated by the Deutsche Bundesbahn and sponsored by the Deutsche Forschungsgemeinschaft. The authors wish to thank the DFG for their valuable support and the DB for the permission to publish the results of this investigation.

REFERENCES

1. Börner, U. and Glaser (1995). *Untersuchung von Oberflächenfehlern an Radscheiben.* Brandenburg-Kirchmöser: Deutsche Bahn AG, Forschungs- und Versuchszentrum Kirchmöser, Versuchsbericht 543-001.

2. Edel, K.-O. (1994). *Einschätzung der Dauerfestigkeit von Eisenbahnvollrädern mit Oberflächendefekten im Radscheibenbereich.* Brandenburg an der Havel: Fachhochschule Brandenburg, Forschungsbericht.

3. Wächter, K. and Näbrich, F. (1986). *Berechnung der Temperaturen und thermischen Spannungen für das klotzgebremste Vollrad für 22,5 t Achsfahrmasse.* Dresden: Hochschule für Verkehrswesen Friedrich List, Untersuchungsbericht.

4. Murakami, Y. (ed.) (1987). *Stress intensity factors handbook* (2 vol.), Oxford: Pergamon Press.

5. Bowie, O. L. (1973). *Solutions of plane crack problems by mapping techniques.* In: Sih, G. C. (ed.) *Mechanics of fracture*, vol. 1. Leyden: Noordhoff International Publishing.

6. Edel, K.-O. and Ottlinger, P. (1990). *Fatigue crack growth characteristics of solid wheels. Rail International* **21**, 15 - 22.

7. Edel, K.-O. and Schaper, M. (1992). *Fracture mechanics fatigue resistance analysis of the crack-damaged tread of overbraked solid railway wheels. Rail International* **23**, 35 - 49.

8. Internationaler Eisenbahnverband (UIC). *Technische Lieferbedingungen für Wagenvollräder.* UIC-Merkblatt 812-3, 2. Ausgabe: 1.1.1963.

9. Internationaler Eisenbahnverband (UIC). *Technische Lieferbedingungen für Vollräder aus gewalztem, unlegiertem Stahl für Triebfahrzeuge und Wagen.* UIC-Merkblatt 812-3, 5. Ausgabe: 1.1.1984.

10. Richard, H. A. (1987). *Praxisnahe Beurteilung von Mixed-Mode-Rißproblemen.* Deutscher Verband für Materialprüfung e.V., Berichte der 19. Sitzung des DVM-Arbeitskreises „Bruchvorgänge", Freiburg, pp. 357 -370.

11. Hänel, B. and Wirthgen, G. (1993). *Rechnerischer Festigkeitsnachweis für Maschinenbauteile.* Dresden: IMA Materialforschung und Anwendungstechnik GmbH, Abschlußbericht.

A NOVEL FATIGUE ANALYSIS APPROACH FOR TUBULAR WELDED JOINTS

E. NIEMI
P. YRJÖLÄ
Department of Mechanical Engineering, Lappeenranta University of Technology,
P.O.Box 20, FIN-53851, Lappeenranta, Finland

ABSTRACT

A novel approach is proposed for fatigue analysis of tubular joints. A local value of nominal stress can be determined using an effective cross-sectional area, or an effective section modulus, derived from known formulae for static strength. The hot spot (or "geometric") stress for fatigue analysis can be calculated using these effective cross-sectional properties with an appropriate stress concentration factor, K_s. The fatigue analysis is then based on a known hot spot S-N curve. Several different joint cases have been studied. It is shown that the factor K_s is similar in the joints and loading cases studied. For initial screening purposes, the use of a constant K_s value of 2.0 appears to be sufficient.

KEYWORDS

Tubular Joints, Fatigue, Nominal Stress, Hot Spot Stress, Novel Approach.

NOMENCLATURE

A	Area of cross-section
A_{eff}	Effective area of cross-section
C	S-N curve constant
E	Elastic modulus
f_y	Yield strength used in design
F	Axial force in a branch member
F_R	Axial force capacity
K_s	Structural stress concentration factor (present method)
m	S-N curve exponent
M	Bending moment in a branch member
M_R	Bending moment capacity
n	Number of tests
N	Number of cycles to failure
N_{exp}	Experimental fatigue life
N_{Ks}	Fatigue life predicted using the present method with $K_s = 2.0$
N_{SCF}	Fatigue life predicted using stress concentration factor, SCF

N_{SNCF}	Fatigue life predicted using strain concentration factor, *SNCF*
R	Stress ratio
SCF	Stress concentration factor (based on finite element analysis)
SNCF	Strain concentration factor, (based on strain measurements)
S_X	Standard deviation of X
W_{eff}	Effective section modulus
X_m	Mean value of X
$\Delta\varepsilon_{nom}$	Nominal strain range
Δ (prefix)	Range
$\Delta\sigma_{e,nom}$	Local nominal stress range resolved using the present method
$\Delta\sigma_{e,hs}$	Hot spot stress range resolved using the present method
$\Delta\sigma_{hs}$	Hot spot stress range at the site of crack initiation
$\Delta\sigma_{nom}$	Nominal stress range based on gross cross-section

INTRODUCTION

Circular (CHS) and rectangular (RHS) hollow sections are widely used in both offshore and onshore structures. Welded joints between chords and branch members are often prone to fatigue failure due to the uneven stiffness distribution and high stress concentrations which exist. The most common fatigue analysis method used in this field is based on hot spot stresses [5]. The stress concentration factor, SCF, varies principally with the width and thickness ratios of the members, as well as other geometric parameters [1]. The conventional stress concentration factor is defined as the ratio of the hot spot stress at the crack initiation site to the nominal stress. The latter is defined as the nominal axial or bending stress in the branch member, depending on the case.

In the conventional hot spot fatigue analysis approach described above, the gross cross section of the loaded member is always used for determination of the nominal stress. This is the reason why large variations exist in the SCF values. The dependence of the stress concentration factor on the geometric parameters of the joint is described by empirical curve-fitted equations. The required data points are usually produced by performing numerous finite element analyses. As this is a tedious task, parametric stress concentration formulae are available only for a few cases. In other cases, the designer must perform finite element analyses, or make strain gauge measurements on prototypes.

In other fields of strength analysis, cases with uneven stress distribution are treated by using effective cross sections rather than gross cross sections. The compression strength of thin-walled structures and the shear lag effect in wide flange beams are examples of such cases. The question arises whether a similar approach could also be used for fatigue analysis of tubular joints. As our goal is to develop an easy-to-use method, the establishment of parametric equations for the ratio A_{eff}/A is out of the question. Another, more easy way, is to derive the effective cross-sectional area or section modulus from the static strength formulae already available for almost all types of connection. Such formulae have been developed either by using the yield line theory, experimental testing [9], or non-linear finite element analyses.

Attempts to relate the fatigue strength to the static strength of the joint have been made, but such methods have not achieved wide acceptance. Objections have been based e.g. on the fact that the static strength depends on the yield strength of the material, but the fatigue strength is independent in the as-welded condition. In the present proposal, this objection is not valid because the yield strength is not included in the calculation procedure. The principle is to calculate a more realistic local nominal stress, hoping that the stress concentration factor will then be practically constant, and easily determined for various types of joint.

The main advantage of the present approach is that it is readily applicable to all types of joint for which static strength can be determined, provided the stress concentration factor can be shown to be relatively constant. The main drawback is that this approach does not differentiate between stresses at different points in the joint. Therefore, it is not suitable for superimposing the effects of simultaneous loads in different members. In such cases, the present approach is suitable for initial screening purposes at the design stage, before more comprehensive analyses are performed, thus saving unnecessary trial and error steps.

FATIGUE LIFE ASSESSMENT METHODS

The fatigue life of tubular joints is assessed using the so-called hot spot approach, which was originally developed for circular hollow section joints used in offshore structures [5]. The life is obtained from the equation:

$$N = C \cdot \left(\Delta \sigma_{hs} \right)^{-m}. \tag{1}$$

Normally, the life is resolved as a lower bound characteristic value, representing a probability of exceedance of about 95 %. In that case, and assuming high residual stresses in the as-welded condition, the following constants are applicable for a material thickness of about 10 mm, corresponding to the Fatigue Class FAT 100 [3]:

$$C = 2 \times 10^{12} \tag{2}$$
$$m = 3.$$

For welds with very smooth reinforcements, FAT 125 with $C = 3.9 \times 10^{12}$ may be considered as a good S-N curve estimate. Similarly, for fillet-welded specimens of thickness below 5 mm, FAT 112 has been chosen.

The principal difference between the conventional hot spot approach, applied to tubular joints, and the present approach is the method of resolution of the hot spot stress range. In the conventional approach the following methods are used:

$$\Delta\sigma_{hs} = \Delta\sigma_{nom} \cdot SCF \tag{3}$$

or

$$\Delta\sigma_{hs} = 1.1E \cdot \Delta\varepsilon_{nom} \cdot SNCF. \tag{4}$$

The factor 1.1 takes into account the average biaxial stress state at the weld toe. The stress and strain concentration factors are determined at the hot spot at the weld toe either by linear or quadratic extrapolation from two or three points. Fig. 1 shows the extrapolation principle in the linear case. Quadratic extrapolation could result in slightly higher estimates for the hot spot stress/strain values than linear extrapolation, depending on the magnitude of the stress/strain gradient.

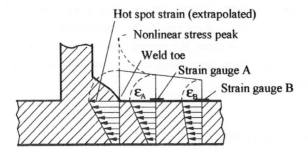

Fig. 1 Linear extrapolation of the strains measured at points A and B.

The present method is based on reduced effective cross-sections. By using an effective area or an effective section modulus, higher local nominal stresses are obtained which are more realistic than those used in the conventional method. Subscripts 'eff' and 'e' are used to denote the quantities specific to the present method. The hot spot stress is resolved using the following formula:

$$\Delta\sigma_{e,hs} = K_s \cdot \Delta\sigma_{e,nom} \tag{5}$$

The local nominal stress is obtained from the following formulae:
- for axial loading

$$\Delta\sigma_{e,\,nom} = \frac{\Delta F}{A_{eff}} \; ; \qquad A_{eff} = \frac{F_R}{f_y} \; \text{or} \; A_{eff} = \frac{F_R}{1.5f_y} \tag{6a}$$

- or for bending loading

$$\Delta\sigma_{e,\,nom} = \frac{\Delta M}{W_{eff}} \; ; \qquad W_{eff} = \frac{M_R}{1.5f_y} . \tag{6b}$$

Thus, the effective area of the cross-section, A_{eff}, and the effective section modulus, W_{eff}, are defined such that when multiplied by $1.0\,f_y$ or $1.5\,f_y$, the static strength of the joint is obtained. The latter case applies when the static strength is governed by plastic bending, either by formation of yield lines in the chord or a plastic hinge in the branch member. As shown in Fig. 2, the pseudo-elastic stress reaches a value of $1.5 f_y$ in such limit states.

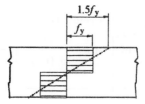

Fig. 2 Pseudo-elastic bending stress at a plastic hinge.

The examples in the following sections provide a more detailed picture of the proposed method.

EXAMPLES

Fig. 3 shows the joint types studied. They have been chosen from the literature on the basis that at least either hot spot stress data (FE analysis or measured) or fatigue testing results are available. Also shown are formulae for A_{eff} or W_{eff}, as applicable. The formulae have been derived using Refs. [2, 7, 10].

Case A comprises axially-loaded RHS X-joints reported by van Wingerde [11]. Fatigue lives, N_{exp}, measured strain concentration factors, *SNCF*, and stress concentration factors, *SCF*, resolved by finite element analyses have been reported. Due to the relatively low breadth ratio, b, failure always occurred in the chord face, governed by plate bending stresses. In the following, 11 specimens are reanalysed; the cases which failed in fatigue testing below 2 million cycles were selected. In this series, the parameters b, g, t (Fig. 3) were constant but the section sizes and load levels were varied.

Case B comprises axially-loaded CHS T-joints. Romeijn [8] reports *SCF* values obtained from parametric formulae for 16 different joint geometries. The fatigue test results are not available. However, the cracks were assumed to develop in the chord, governed by the shell bending stresses. In this case, the predictions made using the conventional hot spot method and the novel method are compared.

Case	Effective cross-sectional properties
A 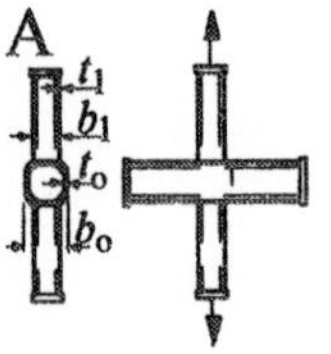 $\beta = b_1/b_o = 0.7$ $\tau = t_1/t_o = 0.63$ $2\gamma = b_o/t_o = 16$	Chord failure: $$A_{\mathrm{eff}} = \frac{t_o^2}{1.5 \cdot (1-\beta)} \cdot \left[2\beta + 4(1-\beta)^2\right]$$
B 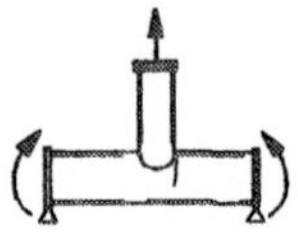	Chord failure: $$A_{\mathrm{eff}} = t_o^2 \cdot \left(1.87 + 9.47\beta^2\right) \gamma^{0.2}$$
C 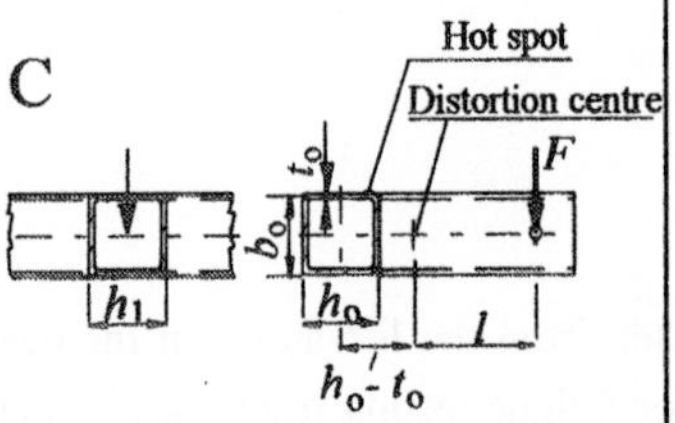	Chord failure: $$W_{\mathrm{eff}} = 1.33 \cdot t_o \cdot \left[h_1 t_o + \sqrt{b_o h_o t_o} \cdot (b_o + h_o)\right]$$
D$_1$ $\beta = 0.6$ 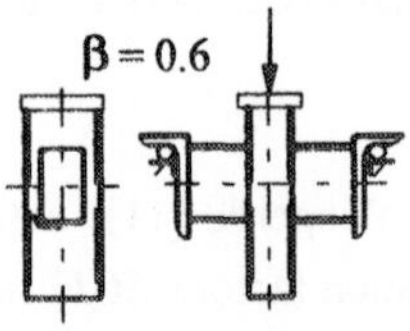	Chord failure: $$W_{\mathrm{eff}} = 0.667\, t_o^2 h_1 \cdot \left[\frac{1}{2h_1/b_o} + \frac{2}{\sqrt{1-\beta}} + \frac{h_1/b_o}{1-\beta}\right]$$
D$_2$ $\beta = 1.0$ 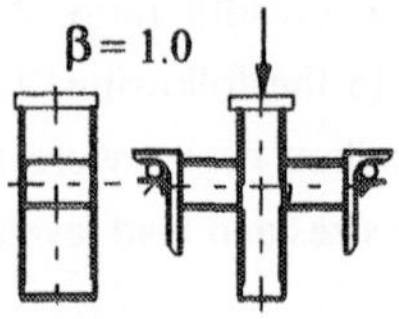	Branch failure: $$W_{\mathrm{eff}} = 0.667\, t_T (h_1 - t_1) \cdot \left(\frac{h_1 - t_1}{2} + \frac{10}{b_o/t_o} \cdot \frac{t_o}{t_1} \cdot b_1\right)$$
E$_1$ 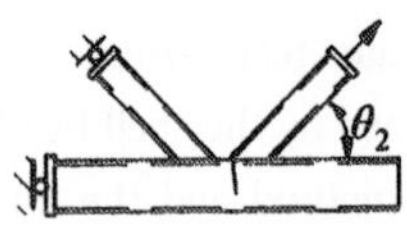	Chord failure: $$A_{\mathrm{eff}} = \frac{5.93 \cdot t_o^2}{\sin\theta_2} \cdot \beta \cdot \gamma^{0.5}$$
E$_2$ 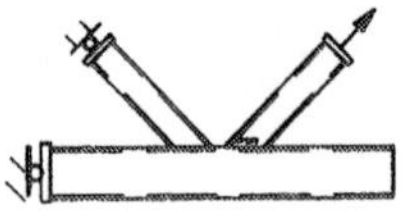	Brace failure: $$A_{\mathrm{eff}} = t_2 \cdot \left[3b_2 - 4t_2 + \frac{10 \cdot t_o^2 b_2}{t_2 b_o}\right]$$

Fig. 3 Joint types studied including the formulae for resolving the effective cross-sectional properties.

Case C comprises RHS T-joints subject to out-of-plane bending. Laitinen has conducted fatigue testing on six joints [4]. The specimens were fabricated from square hollow sections 89 x 89 mm with wall thicknesses of 3.7 and 4.7 mm. The length of the brace member, l, was varied from 131 mm to 313 mm. Longitudinal fatigue cracks developed on the upper face of the chord at the weld toe, governed by the plate bending stresses due to distortion. The welds were rather flat, i.e. only very shallow weld reinforcements were present. All specimens were equipped with one strain gauge, located close to the hot spot in the transverse direction. The stress state was assumed to be uniaxial i.e. the biaxiality effect was not taken into account. Therefore, estimated experimental stress ranges are not very accurate.

Cases D_1 and D_2 comprise RHS X-joints made of austenitic stainless steel, subject to in-plane bending [12]. The wall thickness varied between 3 and 5 mm, and the load level was also varied. Case D_1 represents a rather flexible joint with a breadth ratio $b = 0.6$. In this case, failure occurred in the chord due to plate bending. In the Case D_2 the breadth ratio was $b = 1.0$. This led to branch member failure, governed by membrane stresses at the corners of the cross-section.

Cases E1 and E2 comprise RHS K-joints with gaps between the braces, subject to axial loads in the brace members [6]. In Case E_1, the brace member thickness was equal to the chord thickness, which led to chord failure governed by plate bending stresses in the gap area. Test results were available for stress ratios $R = -1$ and $R = 0.1$. In the former case, 60 % of the compression amplitude was considered as effective assuming moderate residual stresses [3]. In Case E_2, the braces were relatively thin-walled, which led to brace failure, governed by membrane stresses at the corners of the cross-section.

RESULTS

Case A

Only this case is reported here in full. Table 1 shows the results from all 11 specimens.

Statistical analyses

Table 2 shows statistical data for all the cases. Unless stated otherwise, X indicates the difference between the logarithm of experimental life and the logarithm of life predicted using the novel method, i.e. $X = LogN_{exp} - LogN_{Ks}$.

In the proposed approach, a constant structural stress concentration value of $K_s = 2.0$ was used in all cases. If no fatigue test data were available, fatigue lives were predicted using a (gross) nominal stress of 100 MPa.

Table 1. Reanalysed results for 11 axially-loaded X-joints.

Spec-imen	SNCF	SCF	$\Delta\sigma_{nom}$ MPa	A_{eff} mm^2	$\Delta\sigma_{e,nom}$ MPa	N_{SNCF}	N_{SCF}	N_{Ks}	N_{exp}
X1	5.60	6.61	52.9	291	200	57734	46727	31360	105000
X2	7.30	6.51	52.9	291	200	26063	48913	31360	150000
X5	3.80	5.53	44.5	1256	143	310339	134026	85854	280000
X6	3.00	5.53	44.5	1256	143	630700	134026	85854	310000
X7	4.50	5.53	30.0	1256	96	608903	436705	278863	665000
X8	4.15	5.53	30.0	1282	94	776322	436705	296799	1400000
X9	3.05	5.67	54.0	2282	166	336895	69795	54276	100000
X10	3.35	5.67	54.0	2282	166	254250	69795	54276	90000
X11	3.90	5.67	29.4	2282	90	996819	431756	338538	790000
X12	4.28	5.67	29.4	2282	90	754186	431756	338538	330000
X20	13.00	17.32	58.2	276	260	3475	1956	14154	18300

Table 2. Statistical data for the cases studied.

Case	n	Fatigue Class	X_m	S_X	Remarks
			0.3887	0.2264	
A	11	FAT 100	0.0222	0.4411	$X = LogN_{exp} - LogN_{SNCF}$
			0.3268	0.2779	$X = LogN_{exp} - LogN_{SCF}$
B	16	FAT 100	-0.3524	0.1240	$X = LogN_{SCF} - LogN_{Ks}$
C	6	FAT 125	0.2519	0.0695	
			0.4989	0.1018	$X = LogN_{exp} - LogN_{SNCF}$
D_1	18	FAT 125	0.4694	0.1550	
			0.3196	0.2126	$X = LogN_{exp} - LogN_{SNCF}$
D_2	19	FAT 125	0.6348	0.1638	
			0.4676	0.2562	$X = LogN_{exp} - LogN_{SNCF}$
E_1	10	FAT 112	0.7687	0.3326	$R = -1$
	8	FAT 112	0.5202	0.2780	$R = 0.1$
E_2	8	FAT 112	0.2396	0.3502	$R = 0.1$

DISCUSSION

Table 2 shows that the predicted life was usually shorter than the average experimental life, as can be seen from the X_m values, which are greater than 0. This reflects the fact that the S-N curve yields characteristic lower bound values. However, there is a noticeable scatter in the X_m values between different cases. This can be explained by the use of a constant structural stress concentration factor, K_s, and by the somewhat arbitrarily chosen FAT Classes. Many of the specimens were relatively thin-walled, for which a size correction was applied when selecting the FAT Class. The FAT Classes chosen are not ones having wide acceptance.

The use of a constant structural stress concentration factor of 2.0 resulted in acceptable conservative results for most of the cases. Only for the T-joints (Case B), are the results apparently less conservative. By using $K_s = 2.5$, this case would be brought into line with the conventional hot spot prediction. Therefore, it appears to be necessary to study more cases in order to produce a catalogue of stress concentration factors. However, for a certain type of joint and loading condition, a constant structural stress concentration factor appears to be applicable.

Table 2 shows, rather surprisingly, that the standard deviation, S_x, obtained by the novel method is always smaller than that obtained by the conventional method. This fact demonstrates the excellent potential of the novel method in predicting the effect of variations in the geometric parameters of the joints.

The proposed approach is very easy to use. The effective cross-sectional properties were easily determined using well-known formulae for static strength. Special knowledge is required only in concluding whether the static strength is governed by direct stress or by plastic bending. In complicated joints with many potential hot spots, it would be necessary to superimpose the effects of forces and moments in various members. In those cases a variety of K_s factors should be produced using finite element analyses. This would bring this method close to the conventional one. However, it is expected that much simpler K_s relations could be obtained than the conventional parametric formulae.

The local nominal stresses can also be applied directly when the nominal stress approach is used. In that case, the joints could be categorised in Classes between FAT 40 and FAT 64, depending on material thickness and joint configuration.

CONCLUSIONS

Local nominal stresses were resolved for various types of hollow section joints based on effective cross-sectional properties (area and section modulus), derived from known formulae for static strength. Using a structural stress concentration factor of 2.0 and the conventional hot spot S-N curves, good life predictions were obtained. In its present state, the novel method is well suited to preliminary design work, before tedious finite element analyses are performed. The main advantages are simplicity and the availability of the required data for practically all types of tubular joint.

The novel approach shows good potential for development into a general method applicable even to complex tubular joints.

REFERENCES

1.   Efthymiou, M., (1988) Development of SCF Formulae and Generalized Influence Functions for Use in Fatigue Analysis. Shell International Petroleum, Maastschappij B.V.

2.   ENV 1993-1-1/ Amendment 1:1994, Annex K. Eurocode 3: Design of steel structures - Part 1-1: General - General rules and rules for buildings.

3.   IIW doc. XIII-1539-95. Recommendations on Fatigue of Welded Components. Draft for development. Ed. by A. Hobbacher.

4.   Laitinen, P., (1979) On the strength of rectangular hollow section T-joints subject to out-of-plane bending. Masters Thesis, Lappeenranta University of Technology, Department of Mechanical Engineering, Lappeenranta. (In Finnish).

5.   Marshall, P. W., (1992) Design of Welded Tubular Connections. Basis and Use of AWS Code Provisions. *Developments in Civil Engineering*, 37. Elsevier. 412 p.

6.   Noordhoek, C., Wardenier, J. & Dutta, D., (1980) The Fatigue Behaviour of Welded Joints in Square Hollow Sections. *Stevin Report Nr. 6-79-11*. Delft University of Technology, Delft.

7.   Packer, J. A. et al., (1992) Design Guide for Rectangular Hollow Section (RHS) Joints under Predominantly Static Loading. Verlag TÜV Rheinland GmbH, Köln.

8.   Romeijn, A., (1994) Stress and Strain Concentration Factors of Welded Multiplanar Tubular Joints. Ph.D. thesis, Delft University Press, Delft.

9.   Wardenier, J., (1982) Hollow section joints. Ph.D. thesis, Delft University Press, Delft.

10.   Wardenier, J. et al., (1991) Design Guide for Circular Hollow Section (CHS) Joints under Predominantly Static Loading. Verlag TÜV Rheinland GmbH, Köln.

11.   Wingerde, A. M. van, (1992) The Fatigue Behaviour of T- and X-Joints Made of Square Hollow Sections. Ph.D. thesis, Delft University Press, Delft.

12.   Yrjölä, P. & Niemi, E., (1994) Fatigue Tests on X-Joints Made of Austenitic Stainless Steel. Lappeenranta University of Technology, Department of Mechanical Engineering, Lappeenranta. (Unpublished report).

FATIGUE DESIGN OF WELDED THIN SHEET STRUCTURES

J.-L. FAYARD AND A. BIGNONNET

PSA - Peugeot - Citroën - Direction des Recherches et Affaires Scientifiques
Chemin de la malmaison - 91570 Bièvres - France

K. DANG VAN

Laboratoire de Mécanique des Solides - Ecole Polytechnique
91128 Palaiseau - France

ABSTRACT

This work presents the fatigue assessment of an automobile suspension arm made of welded thin sheets, using an original fatigue design method. The approach, which is based on the Hot Spot Stress Concept, defines the use of a unique S-N design curve whatever the geometry of the welded structure and the loading mode. The design stress S, defined as the maximum principal geometrical stress amplitude at the hot spot, is calculated by means of the finite element method (FEM) using thin shell theory. Meshing rules are established for the welded connection and can be applied methodically to any welding situation. The calculation methodology allows the hot spot location and therefore the design stress of any structure to be determined.

KEYWORDS

Design criteria ; Welded structures ; Hot-spot stresses ; Thin shell theory ; FE Method

NOMENCLATURE

N	= number of cycles to fatigue failure.		$R = \dfrac{F_{min}}{F_{max}}$	= load ratio.
S	= design stress.			
σ_L	= actual stress at the hot spot (or local stress).		F_i	= load in the τi direction.
			$e_w^{(p)}$	= weld leg length on the plate side.
σ_{HS}	= geometrical stress at the hot spot.		$e_w^{(a)}$	= weld leg length on the attachment side.
$\Delta\sigma_{HS}$	= σ_{HS} range.		e_a	= thickness of the attachment.
$\dfrac{\Delta\sigma_{HS}}{2}$	= σ_{HS} amplitude.		e_p	= thickness of the plate.
σ_{YS}	= yield stress.		$n_i^{(a)}$	= node number i on the attachment side.
UTS	= ultimate tensile stress.			
F_{min}	= minimum load.		$n_i^{(p)}$	= node number i on the plate side.
F_{max}	= maximum load.			

$E_i^{(a)}$ = shell element number i, perpendicular to the intersection curve of the shell element mean surfaces on the attachment side.

145

$E_i^{(p)}$ = shell element number i, perpendicular to the intersection curve of the shell element mean surfaces on the plate side.

INTRODUCTION

Nowadays, numerous multiaxial fatigue criteria are applied to structures for fatigue life prediction. However, for welded structures in particular, fatigue cracks occur in the weld zones where the stress state is difficult to determine so that existing fatigue criteria cannot be applied. Therefore, the fatigue design of welded structures is usually based on S-N curves [1-3] which relate a design stress S to the number of cycles N characterising failure.

In a previous paper [4], a fatigue design criterion based on a unique S-N curve has been introduced using the hot spot stress concept [5-7]. This design curve is independent of the geometry of the structure and of the loading direction.

The first part of this paper describes the numerical procedure and the experimental basis which lead to the unique design curve. Using this design method, the second part presents the fatigue assessment of a welded automobile suspension arm.

HOT SPOT STRESS CONCEPT AND DESIGN STRESS

Hot spot stress concept

Whenever S-N curves are used for fatigue design, S and N need to be precisely defined. In fatigue, the design stress S must be analysed in the highly stressed zones, called hot spots, where cracks initiate and lead to failure. In the case of continuously-welded thin sheet assemblies, these cracks are usually located at weld toes.

For a welded connection, in the vicinity of these critical areas, the stress can be described either by the *actual stress* or by the *geometrical stress* as illustrated in Fig.1.

The hot spot stress concept assumes that the geometrical stress at the hot spot σ_{HS} is highly related to the stress state which leads to failure, although it does not include local effects. The basic assumption is to consider that continuous welds are of the same type from one component to another and therefore induce comparable local effects (see [4]). In an industrial context, where welding procedures are strictly defined by regulations or in-house codes of practice, this hypothesis is generally true. Therefore, the fatigue design can be based on a structural stress analysis.

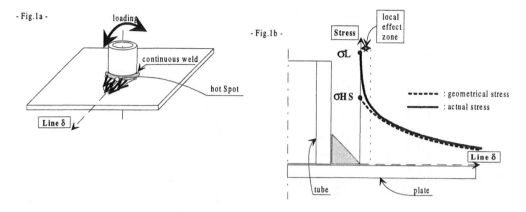

Fig.1 : Illustration of the hot spot stress concept for a tube welded onto a plate and subjected to alternated bending. The following points are indicated : - hot spot location (Fig.1a) and an actual and geometrical stress distribution approaching the weld toe (the weld profile is ideally represented, without any penetration) (Fig.1b).

Numerical analysis

Amongst the numerical analysis methods available for structural calculations, the Finite Element Method allows a fast and clear analysis of the desired information, which are the location of the hot spots and the design stress level. Within the framework of the approach presented in this paper, a static-elastic thin shell finite element calculation is the recommended method to determine the expected design stress [8][9] ; 3D modelling does not provide more accurate information.

In welded connections, crack propagation is considered as the dominant mechanism of fatigue damage. The most significant value for the crack growth is the maximum principal stress. The design stress S is consequently defined as the **maximum principal geometrical stress amplitude** at the hot spot. It can be easily calculated by a numerical analysis.

In a thin shell finite element model, sheets are described by their mean surfaces. However, the outstanding difficulty in using such meshes lies in the modelling of the mean surface intersection. In fact, this zone exhibits 3D behaviour whereas a thin shell model only produces biaxial stresses. Moreover, at the intersection of thin shells, where hot spots commonly appear, the stress gradient can be rather steep so that stress calculations are very sensitive to the mesh size. It is therefore necessary to define a meshing methodology which can be systematically applied to any welded connection. The meshing rules must reproduce :
  (1) The local rigidity induced by the weld size to the structure,
  (2) The stress flow from one sheet to another throughout the weld.
To that purpose, rigid body elements are used to link the two shells.

The size of the elements at the intersection area has been defined such that the geometrical stress is calculated at the weld toe and the weld root without interpolation at nodes, i.e., at the elements' centre of gravity, as shown in Fig.2.

Following these rules, the fatigue design stress of any continuously-welded thin sheet structure can be calculated.

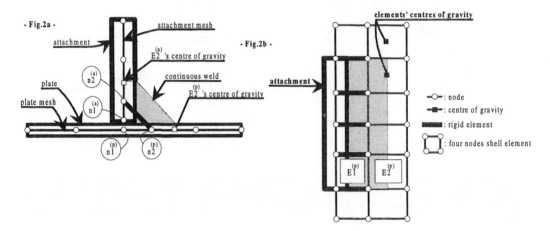

Fig.2. Meshing rules using thin shell elements at the intersection of an attachment welded onto a plate

Experimental basis

The experimental basis concerns automatic MAG welding used for most arc-welding assemblies in the automobile industry. The fatigue curve, or S-N line, is derived from fatigue tests on *elementary structures* (Fig.3), submitted to several kinds of periodic loadings (see [4] for details). These elementary structures are made of commonly used automotive serial brackets welded onto 2 mm thick low strength steel sheets (σ_{YS} =170 MPa) and 2.5 mm thick high strength steel tubes (σ_{YS} =450 MPa). They represent various situations including continuous-welded zones and weld ends with different multiaxial stress states.

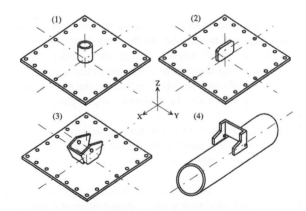

Fig.3. Elementary structures defined for fatigue tests: (1) Tube welded onto a plate. (2) Attachment welded onto a plate. (3) U shape bracket welded onto a plate. (4) U shape bracket welded onto a tube.

A fatigue crack monitoring system was performed with strain gauges at the hot spots. The gauge signal allows the crack depth to be determined. For all testing, a crack depth equal to half of the sheet thickness was defined as the failure criterion N.

Results are shown in a S-N diagram (Fig.4) where the design stress S is calculated by the recommended numerical analysis, and N is the failure criterion. Under these conditions, all the experimental data fall within the same scatter band around a unique S-N curve.

Furthermore, as far as the fatigue crack growth characteristics of all structural steels are roughly the same, one can admit that this S-N curve does not depend on the mechanical properties of the material.

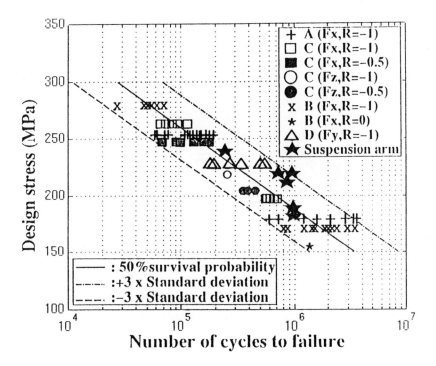

Fig.4 : S-N design curve giving the design stress S versus the number of cycles to failure for continuously-welded structures.

APPLICATION TO AN AUTOMOBILE COMPONENT.

Component, material and loading

The automobile component calculated and tested in fatigue is a front suspension arm (Fig.5). It is made of stamped steel sheets assembled by automatic MAG welding. The rear articulation is a forged steel axle welded onto the two main shells.

Loading is applied through a ball joint and is such that $F(t) = F_m + F_a \sin \omega t$ where F_m and F_a are respectively the mean load and the load amplitude.

The sheets are made from high strength steel, HLE 335 (according to AFNOR std) with UTS=450 MPa and σ_{YS}=335 MPa. Their thickness' are 3 mm for the lower shell, 2.5 mm for the upper shell and 2.5 mm for the attachments.

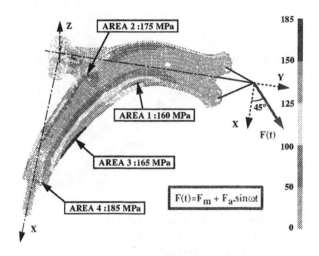

Fig. 5 Maximum principal stress amplitude at the Hot Spots

Numerical model

The numerical model is made of 5808 nodes and 5230 linear integration elements. The representativity of the finite element model boundary conditions has been verified by comparison of the strain calculated at different nodes with the strain gauge measurements. Linear elastic calculation were performed with the MSC/NASTRAN finite element code. The material constants are E = 206,000 MPa and ν = 0.29.

Fatigue life prediction

Prediction. The fatigue criterion requires that the fatigue cycle is described by the amplitude of the applied load. In the present case, it is given by F_a. The calculated value of the design stress displayed on the graphic output, Fig.5, is the maximum principal stress amplitude.

Several hot spots appear in the component with design stresses ranking between 160 and 185 MPa for the applied loading F_a. These results, according to the design curve, indicate that, for this loading, the fatigue life at 50% of survival probability of the component can be estimated respectively between 8×10^5 and 1.75×10^6 cycles. The fatigue cracks will initiate at the described hot spots (area 1, 2, 3 and 4) with a level of probability which can be derived from the design curve.

In a finite element static-elastic calculation, the stress is proportional to the applied load F_a. Therefore, the stress state for any load level proportional to the reference load F_a, can be determined from the calculation performed with F_a. For the four load levels presented in Table 1, with respect to the stresses calculated at the hot spots and according to the S-N curve, the fatigue life at 50% of survival probability can be consequently estimated between two number of cycles.

Experimental validation. Fatigue tests have been performed at four loading levels with a solicitation equal to $F_m \pm F_a$, $F_m \pm 1.1 F_a$, $F_m \pm 1.2 F_a$, and $F_m \pm 1.4 F_a$.

The suspension arm test results presented in Table 1 and in Fig.4 give the number of cycles to the first visible crack. They are in good agreement with the prediction. Cracks initiate at the hot spots which are identified by the calculation, and the fatigue lives correspond to the expected values according to the design curve.

Table 1. Experimental results on the suspension arm.

LOADING	FATIGUE LIFE (number of cycles)		
	Prediction	Test results	(Failure area)
$F_m \pm F_a$	800,000 - 1,750,000	1,000,000	(area 4)
$F_m \pm 1.1 F_a$	450,000 - 1,050,000	1,000,000	(area 3)
$F_m \pm 1.2 F_a$	250,000 - 640,000	880,000	(areas 2 and 4)
$F_m \pm 1.2 F_a$	250,000 - 640,000	960,000	(area 4)
$F_m \pm 1.2 F_a$	250,000 - 640,000	790,000	(area 4)
$F_m \pm 1.4 F_a$	80,000 - 240,000	229,000	(area 2)

CONCLUSION

The originality of the design approach presented for welded structures consists in using a unique fatigue curve whatever the geometry of the structure. Based on the Hot Spot Stress Concept, the design stress is calculated by thin shell modelling using a standardized methodology. Preliminary work on elementary structures allowed a unique S-N curve to be defined.

The efficiency of the approach is shown by an example of an automobile part of complex geometry. The fatigue strength predicted by calculation has been successfully verified by fatigue tests. It is important to note that the use of the Hot Spot approach has the following distinct advantages :
   (1) Cheaper to model using shell rather than 3D elements,
   (2) Faster and easier to model with shell elements.

The final point is most important as this will enable designers, rather than only design analysts (experts), to conduct the assessment. This will ensure its implementation on a high percentage of welded structures and offers numerous possibilities for exploitation.

REFERENCES

1. Commission of the European communities, (1988) *Eurocode 3 : Design of Steel structures, Part 1 - General rules and rules for buildings*, Final draft, Brussels.
2. American Welding Society, (1986) *Structural welding code - Steel*, ANSI/AWS D1.1-86.
3. Department of Energy (1984) *Offshore installations : guidance on design and construction*, HMSO, London.
4. Fayard, JL. , Bignonnet, A. and Dang Van, K. (1996) In: *Fatigue Fract. Engng Mater. Struct.*, Vol. 19, No.6, pp. 723-729.
5. Robert, A. and Radenkovic, D. (1978) In: *Proceedings of the European Offshore Steel Research Seminar*, The Welding Institute, Cambridge.
6. De Back, J. (1987) In: Proceedings of the international conference, *Steel in marine structures*, Elsevier Science Publishers, Amsterdam.
7. Radenkovic, D. (1981) In: *Proceedings of the international conference Steel in marine structures*, Doc. EUR 7347. Pub. IRSID St Germain en Laye.
8. Gerald, J. and Ragout, I. (1992) In: *Proceedings of the international conference Engineering Design in Welded Constructions*, IIW'92, Madrid.
9. ARSEM (1987) *Design guides for offshore structures : welded tubular joints*, Technip, Paris.

FATIGUE OF WELDED JOINTS UNDER COMPLEX LOADING

G R RAZMJOO
Structural Integrity Department
TWI
Cambridge CB1 6AL
United Kingdom

P J TUBBY
Structural Integrity Department
TWI
Cambridge CB1 6AL
United Kingdom

Abstract

The majority of fatigue data for welded joints are for plate specimens under simple uniaxial direct tension, whereas in practice, joints in large structures are frequently subjected to more complex loading. There are very few fatigue test results for such conditions and hence design must currently be based on the published design curves for uniaxial stressing. The objective of the study was to produce fatigue data for joints subjected to complex loading and to evaluate current design recommendations.

This report describes three testing methods developed to produce the following stress conditions:

i)   biaxial bending;
ii)   rotating principal stress direction under bending
iii) in-phase and out-of-phase tension and torsion.

1.   INTRODUCTION

The majority of fatigue tests on welded joints have been conducted on simple plate specimens under uniaxial loading. A statistical analysis of such results for steel joints conducted in 1972[3] led to the development of TWI's fatigue design curves which are now incorporated in many codes and standards, including design codes for bridges[1], pressure vessels[2] and offshore structures[4]. In practice, joints in large structures are invariably subjected to more complex loading which may result in biaxial or triaxial stress conditions, for example, vessel subjected to cyclic internal pressure. In addition the principal stress direction may vary during the loading cycle, as in the web of a girder subjected to a rolling load, for example in a road bridge or crane runway-girder. In contrast to the wealth of experimental data on fatigue of welded joints under unidirectional loading, the data for complex stress states is rather limited. In recent years, sophisticated finite element stress analysis programs and data acquisition systems have enable designers to determine these stresses accurately. The ability to determine stress must be accompanied by the ability to analyse complex stresses in order to predict service life. In view of this, there is now an increasing need to generate data under realistic complex stress states.

The objective of the work presented in this paper was to develop testing methods and obtain constant amplitude fatigue data for joints subjected to complex loading conditions and to evaluate the current design recommendations in the light of the results.

2.   EXPERIMENTAL PROGRAMME

Three testing methods were developed to produce conditions of biaxial bending stresses and rotating principal stress direction in welded plate and tubular specimens. A schematic illustration of the tests is presented in Fig.1. The specimens were made from steel complying with BS 4360 Grade 50D. Chemical analysis and static tensile test results for the material are given in Table 1 and 2 respectively.

Table 1 Chemical analysis of the materials (in wt%)

Test programme	C	S	P	Si	Mn	Ni	Cr
Biaxial bending	0.18	0.008	0.012	0.37	1.31	0.03	0.06
Rotating principal stress in fillet joints	0.11	0.005	0.016	0.39	1.49	0.02	0.03
Tension/torsion	0.16	0.002	0.008	0.38	1.28	0.07	0.09

Table 2 Static tensile properties of the materials

Test programme	Yield strength, MPa	Tensile strength, MPa	Elongation, %	Reduction of area, %
Biaxial bending	396	548	27	66
Rotating principal stress in fillet joints	390	535	29	76
Tension/torsion	398	551	31	77

Welds were made by manual metal arc (MMA) welding using electrodes conforming to BS 639 : 1986 E4333 R (AWS equivalent E6013).

3.   FATIGUE TESTS UNDER BIAXIAL BENDING

3.1.   EXPERIMENTAL WORK

Biaxial cyclic stresses were generated using the testing arrangement in Fig.1a. The specimen comprised a steel disc of 700mm diameter and 12.5mm thickness, simply supported around its periphery and loaded on an annulus of 433mm diameter by a hydraulic actuator. The loading mode is analogous to four point-bending of a beam in that the resulting bending moment is constant within the central area. To facilitate inspection of the tension surface during fatigue testing, observation ports were cut in the specimen support, which comprised a steel cylinder of 750mm diameter and 50mm wall thickness (Fig.1a). Inspection was aided by the use of an inspection lamp and a mirror.

Trial fatigue tests were conducted on disc specimens incorporating fillet welded cover plates. It was found that in the presence of this joint, which has the effect of increasing the local stiffness of the plate, the biaxiality ratio, β, $(=\sigma_2/\sigma_1$, where σ_2 and σ_1 are the principal stresses; $\sigma_1 > \sigma_2$) was significantly affected adjacent to the joint. Whilst in the plain plate remote from the joint, values of β were close to the unity, in the joint vicinity β varied, and was in some cases as low as 0.4, depending on joint geometry.

A series of disc specimens were fatigue tested under biaxial bending stress conditions. The tests were carried out using a servo-hydraulic actuator of 1000kN capacity operating at a frequency of 5Hz at stress ratio, R of zero. A limited number of check tests were also conducted under uniaxial loading conditions for square cover plate joints (Fig.2). These tests were performed in four-point bending using a 50kN actuator at a frequency of 15Hz. Load control was adopted for all tests. The load range was set to give the required principal stress range adjacent to the weld as indicated by foil resistance strain gauges attached to each specimen. In every case the gauge giving the highest principal stress value was used in setting the test load range.

Failure was taken as specimen separation in the uniaxial tests,(under four point bending) and through-thickness cracking in the disc specimens. All specimens were inspected regularly by eye with the aid of soap solution to detect crack initiation and thereafter to monitor crack development.

3.2.   RESULTS AND DISCUSSION

All specimens failed by fatigue crack development at the weld toe, followed by propagation through the plate thickness. As indicated earlier, biaxiality ratio, β adjacent to the failure site varied significantly with specimen design. Figs. 3 and 4 show the relative magnitudes of the principal stresses determined from the strain gauge measurements. The fatigue test results are expressed in the form of S-N curves on logarithmic axes in Fig.5. The data are plotted in terms of the maximum principal stress range in each specimen versus N_1 (crack initiation) and N_2 (final failure). Both the uniaxial and the biaxial results are plotted in the figure. The uniaxial specimens give a reasonably straight line when expressed in terms of N_2 and maximum principal stress range. The uniaxial N_1 data are rather scattered, but this is probably attributable to the experimental difficulties of detecting very small cracks (of the order of 1-2mm surface length). Although subjected to uniaxial loading, β values for these specimens ranged from 0.27 - 0.29. For biaxial loading (under four point bending), β values were very dependent on cover plate geometry, ranging from 0.39 - 0.45 for circular plated (Fig.4) and 0.62 - 0.78 for square plates (Fig.3). All biaxial specimens gave N_2 lives significantly less than expected on the basis of the uniaxial mean line, but there is no consistent trend in terms of the β value. N_2 values for the two specimens with circular cover plates were of the order of five times less the uniaxial mean line, whereas the β values were only marginally higher than those measured in the uniaxial specimens. Clearly, more results are required before realistic conclusions can be drawn.

**4.   FATIGUE TESTS ON NON-LOAD CARRYING FILLET WELDED PLATES UNDER
         ROTATING PRINCIPAL STRESS DIRECTION**

With respect to the case of rotating principal stress direction, there is concern that the method
prescribed in Ref.1 may be unconservative where the direction changes with little of no
change in stress magnitude. Some unpublished results obtained at TWI for beams subjected
to out-of-phase four point bending loads had indicated that fatigue cracking could occur
under these circumstances, although BS 5400 would not predict any fatigue damage. An
experiment was therefore devised to reproduce such conditions as closely as possible and
hence test which design treatment is the most appropriate.

4.1.   EXPERIMENTAL WORK

A sketch of the test specimen and loading arrangement is given in Fig.1(b). The loading was
applied 180° out-of-phase, which produces a stress condition in which the principal stress
direction changes significantly with little change in the principal stress magnitude. The
specimen was of 12.5mm thick plate (500mm x 450mm) with a fillet welded stiffener. As
shown in Figs.1b and 6, the specimen was restrained along one edge by clamping using four
bolts and supported underneath by a 25mm diameter roller.

The fatigue tests were conducted in a specially designed test rig at a frequency of 2Hz under
load control. Equal loads, both nominally at stress ratio, R, of zero were applied out of the
plane of the plate at the two unsupported corners (Fig.1b) causing the plate to bend. Two
separately controlled hydraulic actuators of 30kN capacity were used. Failure was taken
arbitrarily to be the point when a crack of the order of 10mm long could be detected visually.
Subsequent crack propagation was also monitored in some of the specimens.

4.2.   RESULTS AND DISCUSSION

The fatigue test conditions covered a number of principal stress ranges. Fatigue failures
occurred under these loading conditions. Fatigue cracks were detected on the weld cap at the
middle of the fillet weld as well as the weld toe.

The strain measurements were recorded (using rectangular rosette strain gauges 5mm away
from the weld toe, Fig.6) at various points in the load cycle: full load at position A, equal
loads at A and B, full load at B (Fig.6). The greatest variation in the maximum principal
stress range, $\Delta\sigma_1$, and the angle of rotation, θ, were measured at the failure locations. The
results showed that the angle of rotation at the location of failure ranged from 33° to 41° in
different tests. Fatigue assessment, following the recommendations in BS 5400[1] was made by
taking the greatest variation in the maximum principal stress in the cycle as the appropriate
stress (since the angle of rotation was less than 45°). The BS 5500 fatigue assessment was
carried out in accordance with Enquiry Case 79[2]. The strain measurements made at the
position nearest to the location of failure, for the loading conditions at A and B were used to
evaluate the maximum stress range $\Delta\sigma_{max}$. Fatigue assessment was then made using the
appropriate design curve for the specimen, Class F. Results are expressed in the form of an S-
N curve on logarithmic axes in Fig.7.

Figure 7 shows considerable scatter in the results when expressed in terms of the maximum
principal stress range according to the BS 5400 recommendations. This implies that the

principal stress range may not be the appropriate equivalent stress criterion. The figure indicates that the fatigue life predictions based on this code could be unsafe. It also shows that the procedure in BS 5500 Enquiry Case 79 gave a much better fit to the data, but may still result in unsafe fatigue life predications in the case of rotating principal stress.

5.   FATIGUE TESTS UNDER COMBINED TORSION AND TENSION

The case of combined loading of welded tubular joints has been the subject of some interest in the recent years. Amongst the more comprehensive studies are those of Sonsino[6] at LBF and Lawrence et al [5,7] at the University of Illinois, both of which considered the fatigue performance of tube-to-flange joints under combined bending and torsion. The current work at TWI considers the case of tube-to-flange fillet welded joints under combined tension and torsion. Some of the preliminary results are presented in this paper.

5.1.   EXPERIMENTAL WORK

A series of tube-to-flange fillet welded joints were subjected to pure tension, in-phase and out-of-phase tension and torsion. Phase displacement of 90° was selected for the out-of-phase loading. The combined loading tests were carried out under a range of nominal biaxiality ratio $\lambda = \tau/\sigma$, where τ and σ are the shear and tensile stresses respectively.

The specimen geometry is shown in Fig.8. (The tube was bored internally at the ends to reduce the wall thickness in order to achieve the required stress ranges). The specimens were instrumented using rosette strain gauges attached 3mm away from the weld toes. The tests were conducted using a tension/torsion servo-hydraulic test machine with tension and torsion load cell capacities of 100kN and 6kNm respectively.

5.2.   RESULTS AND DISCUSSION

All the specimens failed by fatigue crack initiation at the weld toe, and predominately at the location of the weld stop/start. In the case of combined loading the fatigue crack initiated at the weld toe and propagated away from the weld into the tube material. The cracking direction was perpendicular to the maximum principal stress for the in-phase loading. In the case of out-of-phase loading multiple crack initiation occurred and the crack propagated in a jagged mode.

The preliminary test results are presented in Fig.9. For the case of in-phase loading, the results suggest that the use of principal stress range as the design parameter is satisfactory, albeit conservative. The most striking feature of the results is that fatigue life under out-of-phase loading is much lower than that of in-phase loading. The current results suggest that when expressed in terms of principal stress range (Fig.9) the fatigue endurance under out-of-phase loading is reduced by approximately a factor of 10. The figure also shows that some of the results fall below the design line. The implication is that unlike in the case of in-phase loading, the principal stress range may not be the appropriate design parameter for the case of out of phase loading.

6.   CONCLUSIONS

Testing arrangements were developed to allow tests to be conducted under biaxial bending and rotating principal stress conditions for non-load carrying fillet welded joints as well as load carrying fillet welded tube-to-flange joints in steel. Preliminary results allow the following tentative conclusions:

i)   Under applied equibiaxial bending condition the biaxiality ratio adjacent to a welded joint is strongly dependent on joint geometry. Typical values measured in this programme were 0.39 to 0.45 for circular cover plates and 0.62 to 0.78 for square cover plates.

ii)   Cover plates under biaxial bending gave lower fatigue endurances than those under uniaxial loading when expressed in terms of principal stress range.

iii)   Out-of-phase torsion and tension loading can reduce the fatigue life of fillet welded joints in comparison to that under in-phase loading. A life reduction of the order of magnitude life reduction was noted when expressed in terms of principal stress range.

iv)   The use of maximum principal stress range as the fatigue damage parameter resulted in satisfactory fatigue life prediction for the case of in-phase loading but showed that it could be unsafe for the case of out-of-phase tension and torsion loading.

v)   The fatigue design methods in the current design code[1] may not adequately predict the fatigue life under the condition of rotating principal stress.

7.   ACKNOWLEDGEMENT

The work was funded jointly by Industrial Members of TWI and by the Manufacturing and Technology Division of the United Kingdom Department of Trade and Industry.

8.   REFERENCES

1.   BS 5400 - Steel, concrete and composite bridges, Part 10, Code of Practice for Fatigue. British Standards Institution, London, 1980.

2.   BS 5500 : 1991 - Specifications for unfired fusion welded pressure vessels. Enquiry Case 79. Assessment of vessels subject to fatigue: alternative approach to method in Appendix C. British Standards Institution, London.

3.   Gurney, T R and Maddox, S J. 'A re-analysis of fatigue data for welded joints in steel.' *Welding Research International*, vol.3, no.4., pp.1-54.

4.   'Offshore Installations: Guidance on design, construction and certification', Department of Energy, HMSO, Fourth Edition, 1990.

5.   Siljander, A, Kurath, P and Lawrence, F V Jr, 'Multiaxial non-proportional fatigue of weldments. *EIS 90*, England.

6.   Sonsino, C M. 'Fatigue of welded joints under multiaxial in-phase and out-of-phase local strains and stresses.' *Fourth International Conference on Biaxial/Multiaxial Fatigue*, Paris, 1994.

7.   Yung, J Y and Lawrence, F V Jr. 'Predicting the fatigue life of welds under combined bending and torsion'. *Biaxial and Multiaxial Fatigue*. Edited by M W Brown and K J Miller, Mechanical Engineering Publications, London, 1989. pp.53-69.

G. R. Razmjoo and P. J. Tubby

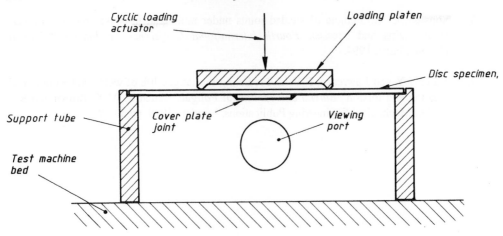

a) Biaxial loading

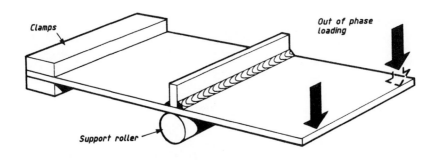

b) Rotating principal stress direction

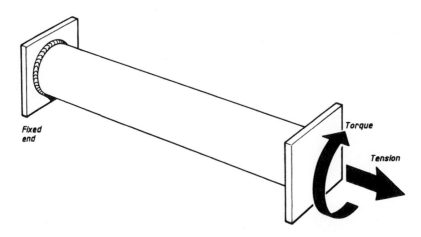

c) Combined torsion and tension.

Fig.1 Schematic illustration of testing arrangements to produce complex loading

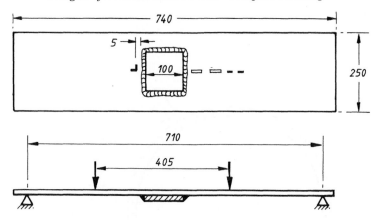

Fig.2 Uniaxial fatigue test arrangement.

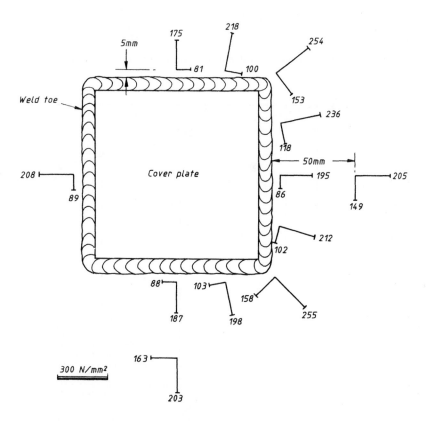

Fig.3 Measured principal stress distribution for square cover plate joints under biaxial loading.

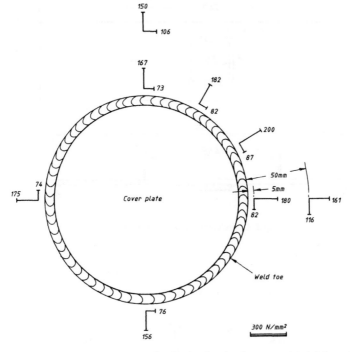

Fig.4 Measured principal stress distribution for circular cover plate joints under biaxial loading.

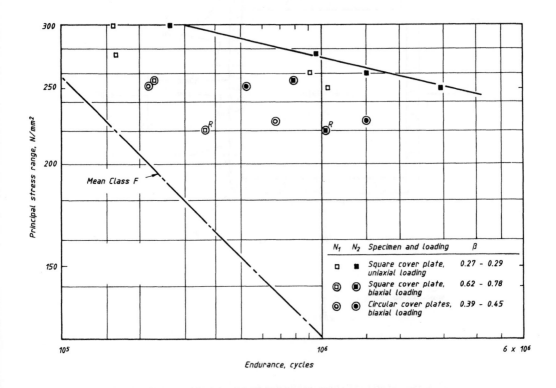

Fig.5 Fatigue test results for cover plate joints.

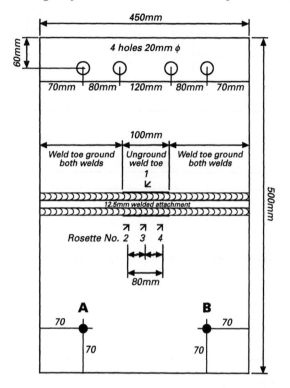

Fig.6 The specimen geometry for non-load-carrying joints under rotating
principal stress direction.

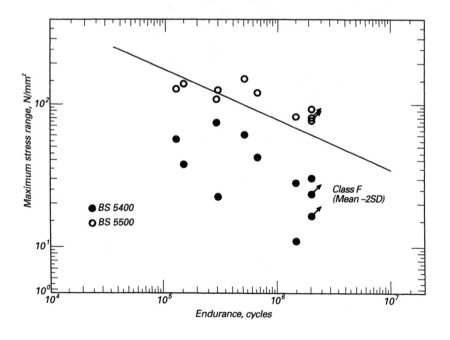

Fig.7 Fatigue test results for non-load-carrying joints under rotating principal
stress direction.

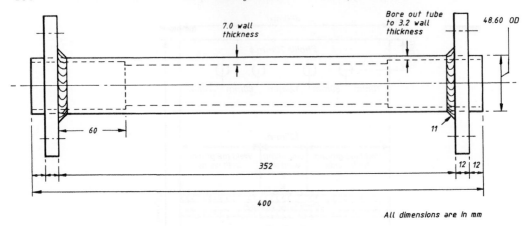

Fig.8 The specimen for tests under combined torsion and tension.

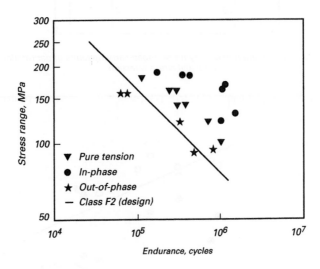

Fig.9 Fatigue test results for joints under combined torsion and tension (in terms of maximum principal stress).

EFFECTIVENESS OF IMPROVEMENT METHODS FOR WELDED CONNECTIONS SUBJECTED TO VARIABLE AMPLITUDE LOADING

V. Dubois, Swiss Federal Aircraft Factory, CH-6032 Emmen, Switzerland
M. A. Hirt, Swiss Federal Institute of Technology, CH-1015 Lausanne, Switzerland

ABSTRACT

The fatigue resistance of welded details can be increased by improvement methods. Residual stress methods have been shown to be effective under constant amplitude loading whereas less is known about their effectiveness under variable amplitude loading. This paper summarises the results of research on the fatigue behaviour of longitudinal welds treated with improvement methods and under variable amplitude loading.

KEY WORDS

Variable amplitude loading, welded connection, residual stresses, needle peening, fatigue life improvement, crack propagation

INTRODUCTION

Welded structures are often subject to fatigue problems. Improvement methods that act on the residual stress field can be considered to enhance fatigue strength. Despite a certain amount of research in this area [1 - 4] there are no design recommendations for the use of improvement methods.

The first part of this paper summarises the results of fatigue tests under variable amplitude loading on specimens with longitudinal attachments. The second part describes a parametric study undertaken with the use of a computer model developed to simulate fatigue crack propagation. Finally, modified S-N curves are proposed for the rational use of improvement methods on details subjected to variable amplitude loading.

LABORATORY TESTING

Fatigue testing was conducted on specimens made from high strength steel of grade Fe E 355. Each specimen consisted of a 1 metre long base plate with 200 mm long gusset plates, fillet welded onto each side (Fig. 1). As-welded stress-relieved specimens, improved by needle peening, were tested. Details of this improvement method can be found in [2].

165

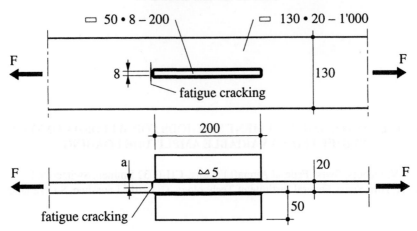

Fig. 1 Geometry of the test specimens (dimensions in millimetres).

Two measuring systems were installed on each specimen. One system allows monitoring of crack propagation during the test and is based on the potential drop method [5, 6]. The other system allows monitoring of crack-opening stress [7]. This stress is defined as the stress at which the crack is open and thus susceptible to propagation.

Crack-opening stress indicates the effects of residual stress and plasticity at the crack tip. Two stress spectra (Fig. 2) were chosen for studying the influence of the crack-opening stress. The characteristics of the two spectra were determined on the basis of a parametric study [8].

TEST RESULTS

Crack Propagation

Figure 3 shows the results of crack propagation for both spectra, A (constant maximum stress) and B (constant minimum stress). In the case of spectrum A, crack propagation is slightly faster in unimproved details than in needle peened details. However, the difference is not significant when dispersion is taken into account. For spectrum B, crack propagation for improved specimens is much slower than for unimproved specimens. This illustrates the effect of the value of the constant minimum stress value of the small stress ranges, $\Delta\sigma_s$, in the spectrum.

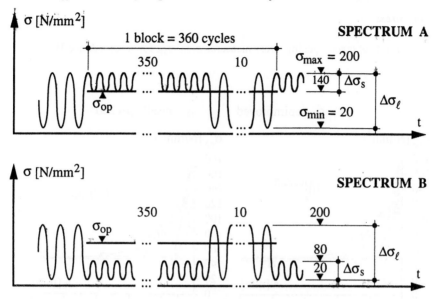

Fig. 2. The two spectra chosen for variable amplitude loading: spectrum A (constant maximum stress) and spectrum B (constant minimum stress).

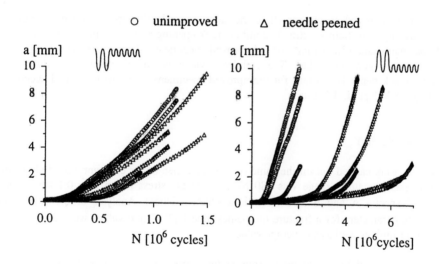

Fig. 3. Crack depth (a) as a function of number of cycles (N) for spectra A and B.

Crack-opening Stress

Figure 4 shows results of crack opening stress measurements, σ_{op}, for spectra A and B.

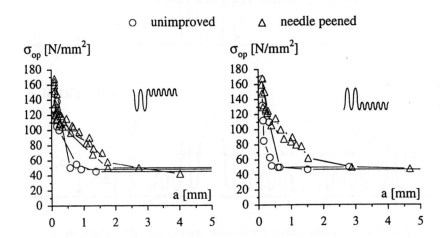

Fig. 4. Crack-opening stress (σ_{op}) as a function of crack depth (a) for spectra A and B.

Unimproved and improved specimens show the same general relationship between crack opening stress and crack depth for both spectra. The initial crack opening stress is quite high (160 N/mm^2 for improved specimens and 130 N/mm^2 for unimproved specimens). This stress decreases rapidly and then stabilises at a constant value. The stabilised value is approximately 50 N/mm^2 and the corresponding crack depth is 0.5 mm for unimproved specimens. For needle peened specimens, the corresponding crack depth is 1.7 mm.

PARAMETRIC STUDY

Previously, tests were carried out on the same type of specimen under constant amplitude loading [2, 9]. For these tests, the stress range was 180 N/mm^2 and the stress ration R = 0.1 (the large stress ranges, $\Delta\sigma_l$, of spectra A and B also have these characteristics). Analysis of the results allowed definition of the parameters for a fracture mechanics model [9]. Three values were defined: an upper limit and a lower limit as well as an average value.

A crack propagation model was developed from the test results, including those found in the literature [9]. This model allowed consideration of: applied loads, residual stresses and plasticity at the crack tip.

The model was then used for a parametric study with the following assumptions:
- an average distribution of residual stresses based on the results reported in [10],
- initial crack depth 0.1 mm, final crack depth 10 mm,
- angle between tangent to fillet weld and base plate equal to 45°.

The tests have shown that the needle peening acts on the distribution of residual stresses at the weld toe. This same effect was shown by simulations for the same spectra (Fig. 5). Residual stresses have less effect in the case of Spectrum A (constant maximum stress) where the minimum stress of all small stress ranges, $\Delta\sigma_s$, is above the crack opening stress. For other spectra, residual stresses have more effect.

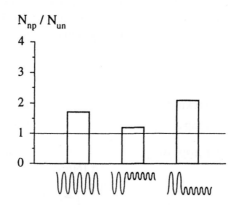

Fig. 5. Ratio of number of cycles to failure for needle peened specimens to that for unimproved specimens (N_{np} / N_{un}) and for different load spectra.

The fatigue behaviour of needle peened and unimproved specimens under variable amplitude was further investigated with spectra consisting of various shapes of stress block. The characteristics of the spectra were specifically chosen to study the influences of the following:
• the position of the stress block with respect to the CAFL (Constant Amplitude Fatigue Limit),
• the shape of the stress blocks,
• constant minimum or constant maximum stress.

Full details of the results [9] can be summarised as follows. The value of the constant stress in the spectrum determines whether an improvement method can increase fatigue life. A stress spectrum with constant maximum stress renders an improvement method less effective - in fact the entire effectiveness may be lost. The effectiveness of an improvement method is greatest when the applied spectrum has many small stress ranges and their maximum stress remains below the crack opening stress.

RESISTANCE CURVES (S-N CURVES)

Improvement methods allow an increase in the number of cycles to failure of a welded detail but their application to real structures should follow specific rules. The use of S-N curves allows consideration of the same parameters as used for the parametric study.

Upper and Lower Limits

The number of cycles to failure under constant amplitude loading was simulated for each stress range using the crack propagation model with R = 0.1 and with two limits of the parameters of the fracture mechanics model, as reported in the parametric study portion of this paper.

For both improved and unimproved specimens, the results lie within the dispersion limits given by the fracture mechanics model (Fig. 6). The average slope constant of the two straight lines is 3 for unimproved and 4 for the needle peened specimens. The calculated CAFL is much higher than that of the code [11]. This is explained by the fact that the simulations were made using R = 0.1, whereas the code takes into account test results with all R ratios.

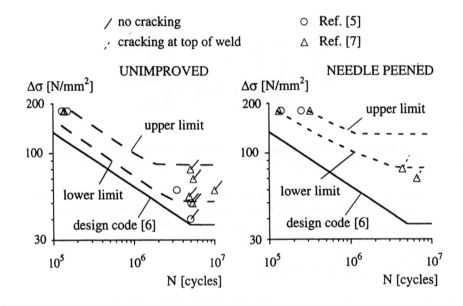

Fig. 6. Simulated S-N curves of unimproved and needle peened
specimens compared to the design code [11] and test results.

Effect of the Type of Improvement Method

Additional S-N curves were simulated for different types of residual stress improvement methods. These simulations were carried out using the same hypotheses as previously. Only the lower limit of the fracture mechanics model is shown in Fig. 7. These curves implicitly contain the effects of the distribution of residual stresses.

The simulated S-N curves of Fig. 7 clearly show that the most effective treatment is hammer peening, followed by needle peening and shot peening. This corresponds to the experimental findings of previous research [2, 12, 13]. The value of the slope constant increases with increasing effectiveness - as does the CAFL. The same phenomena have been observed for other geometrical arrangements [3, 12]. Hammer peening is the treatment that modifies the distribution of residual stresses the most, in value and in depth [10]. This treatment is best for high stress ranges.

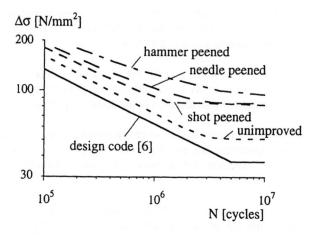

Fig. 7. Simulated S-N curves for different improvement methods.

Effect of Maximum Stress

In order to observe the effect of the maximum stress on the S-N curve, the curves were simulated by calculating the number of cycles to failure for each level of stress, but keeping the maximum stress constant. Several values were chosen for the maximum stress (80, 100, 150, 200 N/mm²). The results are shown in Figure 8. This figure also shows the results of simulations considering a constant minimum stress equal to zero.

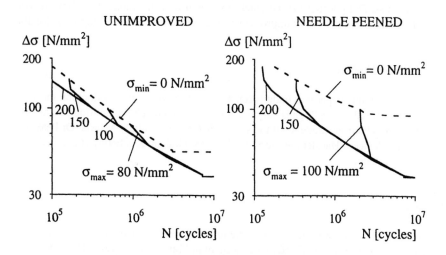

Fig. 8. Simulated S-N curves for different values of maximum
stress for unimproved and needle peened specimens.

There is not a significant difference between the curves for unimproved specimens. All curves commence at an upper limit represented by the simulated S-N curve for zero minimum stress and join the simulated curve for maximum stress equal to 200 N/mm^2.

Curves for needle peened specimens differ greatly. For all values of maximum constant stress, the curves descend rapidly and join the lower curve obtained for 200 N/mm^2. The effect of constant maximum stress must be an important consideration in the design of peened joints.

These simulations confirm the test results in [14] where an increase in average stress did not influence the S-N curve for untreated specimens but actually decreased the fatigue resistance of peened joints. Other results, such as in [15] confirm that in tests where the maximum stress was constant, all beneficial effects of the treatment were lost.

To conclude, the use of an S-N curve obtained from constant maximum stress covers all test results. For unimproved specimens the use of such a curve is justified. If this curve is used for peened specimens, the effectiveness of the treatment is not taken into account. The curve should consider the stress spectrum, in particular the effect of the maximum stress.

CONCLUSIONS

This study has shown that the stress range and the value of the maximum stress affect crack propagation under variable amplitude loading. These two values must be considered in the fatigue design of details that have been treated with residual stress methods in order to account for the effectiveness of the improvement.

The crack propagation model has enabled three significant parameters to be taken into account: applied stress, residual stress, and plasticity at the crack tip. The use of this model has allowed S-N curves to be proposed that allow rational use of improvement methods under variable amplitude loading.

ACKNOWLEDGEMENTS

The work described in this paper was undertaken as part of the first author's doctoral thesis at ICOM and was partly sponsored by the "Fonds national suisse de la recherche scientifique". The authors would like to express their gratitude to all the staff at ICOM, particularly Messrs. Dr Peter Kunz, Prof. Eugen Bruehwiler, Vincent Haesler and Eerik Peeker. Mr Graham Kimberley is thanked for the translation of this article.

REFERENCES

1. Bignonnet, A., Improving the fatigue strength of welded steel structures, Steel in Marine structures. Proceedings of the Conference on Steel in Marine Structures (SIMS PS 4). Amsterdam: Elsevier, 1987.
2. Bremen, U., Amélioration du comportement à la fatigue d'assemblages soudés: Etude et modélisation de l'effet de contraintes résiduelles. Thèse no. 787, Ecole Polytechnique Fédérale de Lausanne. Lausanne: EPFL, 1989.

3. Haagensen, P. J., Weld improvement methods for increased fatigue strength. Engineering design in welded constructions. Proceedings of the International Conference of the International Institute of Welding. Oxford, Pergamon Press, 1992, pp. 73-92.

4. Smith, I. F. C., Dubois, V., and Bremen, U., Peening methods for improvement of fatigue strength. Paper presented at the meeting of IIW Commission XIII, WG2 (Doc. IIW WG2-2-90). Lyon, 1990.

5. Dubois, V., et Bremen, U., Détermination de la profondeur de fissures de fatigue à l'aide d'un champ électrique. Lausanne: Ecole Polytechnique Fédérale de Lausanne, publication ICOM 202, 1988.

6. Smith, I. F. C., and Smith R. A., Measuring fatigue cracks in fillet welded joints. International Journal of Fatigue, Vol. 4, No. 1, 1982, pp. 41-45.

7. Bremen, U. Détermination de la contrainte d'ouverture de fissures de fatigue au pied de cordons de soudure. Lausanne: Ecole Polytechnique Fédérale de Lausanne, publication ICOM 200, 1988.

8. Dubois, V., Résistance à la fatigue des assemblages soudés traités: étude pour l'établissement d'un spectre d'amplitude variables et du programme d'essais. Lausanne: Ecole Polytechnique Fédérale de Lausanne, publication ICOM 284, 1993.

9. Dubois, V., Fatigue de détails soudés traités sous sollicitations d'amplitude variable, Thèse no. 1260, Ecole Polytechnique Fédérale de Lausanne. Lausanne: EPFL, 1994.

10. Bremen, U. Détermination des contraintes résiduelles au pied de cordons de soudre bruts ou traités. Lausanne: Ecole Polytechnique Fédérale de Lausanne, publication ICOM 188, 1987.

11. ECCS/TC6 Recommendations for the fatigue design of steel structures. ECCS European convention for construction steel work, Brussels, 1985 (ECCS No. 43).

12. Gurney, T. R., Fatigue of welded structures, 2nd ed. Cambridge, Cambridge University Press, 1979.

13. Maddox, S. J., Fatigue strength of welded structures, 2nd ed., Abington Publishing, 1991.

14. Booth, G. S., The fatigue life of ground or peened fillet welded steel joints - the effect of mean stress, Metal Construction, 1981, Vol. 13, No. 2, pp. 112 - 115.

15. Ohta, S., et al. Methods of improving the fatigue strength of welded joints by various toe treatments. Sub-committee of improving the fatigue strength of welded joints by various weld toe treatments, Society of Steel Construction of Japan. Paper XIII-1289-88 presented at the annual assembly of the International Institute of Welding, 1988.

INFLUENCE OF LIFE IMPROVEMENT TECHNIQUES ON DIFFERENT STEEL GRADES UNDER FATIGUE LOADING

LUIS LOPEZ MARTINEZ[1,2] and ANDERS F. BLOM[2,3]

[1] Application Research and Development, SSAB Oxelösund AB, S-613 80 Oxelösund, Sweden
[2] The Royal Institute of Technology, 100 44 Stockholm and [3]The Aeronautical Research Inst. of Sweden

ABSTRACT

Different fatigue life improvement techniques were applied to test specimens with a non-load carrying fillet weld. These test specimens were then tested under both constant amplitude and variable amplitude fatigue loading.

For the as-welded specimens fatigue lives have been related to weld flaws, mainly cold laps, detected at the fracture surfaces. The size and distribution of cold laps, which are a direct function of plate surface quality, have shown a consistent influence on fatigue lives.

TIG-dressing of weld toe region has shown a fatigue life improvement ranging from 60% to 120% of applied stress range. Steel grades with higher yield strength have shown larger improvement than ordinary structural steels. The influence of blast-cleaning on fatigue properties is addressed. Mechanisms of improvement are discussed and X-ray diffraction measurements of initial residual stress distributions and relaxation due to fatigue loading are used to explain part of the life extension obtained. The paper also includes fractographic studies of the broken fatigue specimens.

KEYWORDS

Improvement Methods, Random Loading, High-strength steel, Residual Stress Relaxation, Fatigue Testing

INTRODUCTION

The increased interest in a higher pay-load in machinery, such as earth moving equipment, has focused attention on a more widespread use of high-strength steels. High-strength steel should allows, at least under static loading, an increase in design stresses so the plate thickness can be reduced.

However, the design stresses are always limited by the fatigue strength of welded joints subjected to fluctuating loads. It is a well known fact that this fatigue strength is more or less independent of parent plate mechanical properties.

175

It is then of main concern to improve the fatigue resistance of welded joints so that the material can be used more effective. This improvement in the fatigue strength, should enable the design stresses to be increased.

There are four major reasons for the reduced fatigue strength in welded joints:

The global stress concentration:
> The change in plate shape always gives a concentration of the force lines in such a way that the more pronounced the change in section, the higher the stress concentration.

The residual stresses:
> Every weld process produces tension residual stresses which are close to the level of yield strength of the filler metal. These residual stresses reach their highest level at the weld toe/fusion line.

Weld flaws:
> Weld toes commonly exhibit such flaws as lack of fusion, undercut, intrusions and cold laps. These are always at the same location as the highest stress concentration and highest residual stress.

Heat affected zone:
> After welding the microstructure will be changed locally and a so-called heat affected zone exists. Embrittlement may result, but normally the above three factors are of much larger concern.

By applying fatigue life improvement techniques to welded joints one or more of these problems can be almost completely remedied. One of the improvement methods presented in this paper, TIG-dressing, favourably affects all of the causes mentioned above. Other methods, as shot peening or hammer peening, are mainly intented to induce compressive residual stresses at weld toe.

Since the fatigue life improvement techniques are sensitive to the type of fatigue load, constant- or variable-amplitude, it is of paramount importance to understand the reciprocal influence between some selected fatigue life improvement technique and a specific load sequence.

Ideas for improving the fatigue lives of welded joints have been the subject of many IIW documents [1]. These focused attention on the weld toe in steels, as a prime site for fatigue cracks. Several fatigue design standards, namely BS 7608:1993 [2] and IIW Doc XIII-1539-94 [3], mention the possibility of increasing the design stress level if a remedial treatment of the weld toe region by controlled machining or grinding is implemented. In the case of BS it is stated that such treatment leads to an increase in the allowed design stress range. The SN-curve can be then improved in strength by 30%. This is equivalent to a factor of 2.2 on life. Consequently the BS presents a clear description of the treatment that should be carried out regarding tools, machining deep and posterior control. The IIW document consider also the application of fatigue life improvement techniques. The IIW document goes further than BS and point out a serie of techniques to improve fatigue resistance. Nevertheless, the IIW document point out the necessity of fatigue test in order to verify the actual procedure for the specific stress range[1] of interest and the fatigue life improvement used.

One of the main drawbacks in the application of fatigue life improvement methods is the variation in the quality of the different process as well as the uncertainty of the degree of improvement.

To date it is very scarse information, if any, on fatigue life improvement methods combined with variable amplitude fatigue loading. Most of the information available on this subject refers to constant amplitude loading. The main task of the present paper is to illustrate the effect of different spectrum parameters on the level of improvement obtained by using some of these methods on different steel grades.

As some of these fatigue life improvement methods are based on the degree of compressive residual stresses that can be induced on the weld toe region, another task is to elucidate the influence of parent plate yield strength when the applied improvement method is based on the level of compressive residual stresses induced.

In the former Nordic Project it has been documented [4] that weld defects are often the site for fatigue cracks to start. The application of fatigue life improvement techniques change these coditions. Consequently this paper includes some fracture surface analysis in order to investigate if these defects are still the origin of fatigue cracks.

EXPERIMENTAL PROCEDURES

Materials and test specimens

The mechanical properties of the steels used in this investigation are presented in Table 1. These steel grades include two (HSLA-steels) cold forming steels DOMEX 350YP and DOMEX 590XPE (extra high strength steel). The present investigation also includes one quenched and tempered steel grade, namely WELDOX 700. This is a so called martensitic steel, extra high-strength structural plate with impact guaranteed toughness at temperatures down to -40° C (-76°F) The chemical composition and carbon equivalent CE_{IIW} types, for the different steels grades are presented in Table 2. The definition of carbon equivalent. CE_{IIW}, is as follows: CE = C + Mn/6 + (Cr+Mo+V) / 5 + (Cu+Ni) / 15.

Table 1 Mechanical properties of the steel grades used in this investigation

Steel	R_{eH} [MPa]	R_m [MPa]	A_s [%]	Impact energy CV(-40°C, J)
DOMEX 350YP	398	503	34	-----
DOMEX 590XPE	615	747	31	27
WELDOX 700	780	850	12	L: 40 T: 27

---

[1]In this case the stress range includes the effect of the stress ratio on compressive residual stresses.

Table 2 Chemical composition and carbon equivalent (CE) for steel grades used in this investigation

Steel	C	Si	Mn	P	S	Al	Nb	V	CE
DOMEX 350	.058	0.02	0.62	.009	0.01	.042	.014	----	0.18
DOMEX 590	.090	0.21	1.63	0.11	0.02	.030	.024	0.15	0.41
WELDOX 700	0.15	0.44	1.32	.012	.002	.099	.060	.060	0.37

Fatigue test specimens of the type illustrated in Figure 1, were used in all the tested series. This is a well documented specimen containing a non-load carrying fillet weld used in several investigations [5]-[8]. Since the thickness and shape of the present specimen is the same as in some of the previous references, a comparison of fatigue test results can be easily done.

WELD PROCEDURE AND FATIGUE LIFE IMPROVEMENT METHODS

The weld procedure is described in Table 3 for the three steel grades tested. No pre-heat treatment has been used. The order of weld passes has been chosen in such a way that the shape and magnitude of the residual stresses should be comparable for all the steel grades involved.

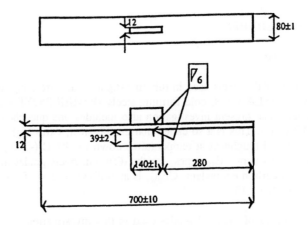

Fig. 1. Test specimen. Dimensions in mm.

The weld on the sides of the stiffener as well as at the corners has been produced in an alternanting diagonal sequence in order to limit the interpass temperature ($<250°$ C). The same welder have manufactured all the test specimens for all the test series at SSAB Oxelösund AB Laboratories, in order to keep the scatter of weld quality under a controlled level. No root treatment or weld preparation have been done for series not including hammer peening improvement. For this last series it has been necessary to achieve a full penetration weld to obtain the same failure mode as for the rest of test series. As it is demonstrated under the heading Residual Stress Measurements, this does not affect the general shape or initial residual stress distribution for the considered fatigue test specimens.

As the same filler metal have been used for the different steel grades tested, a degree of mismatching is present. The yield and tensile strengths for the used filler metal are 420 MPa and 500-570 MPa respectively. This means a degree of overmatching for DOMEX 350 as a degree of under matching for the WELDOX grade. This aspect should have some influence on the max. level for weld induced residual stresses.

Table 3 Weld procedure for the specimens tested in this investigation

Steel	Welding process	Consuma-bles	Electrode diam. [mm]	Current DC+ [A]	Voltage [V]	Heat imput [kJ/mm]
DOMEX 350 DOMEX 590 WELDOX 700	MAG	PZ 6130 Mison 25	1.6	185	23.5	1.5

The parameters for TIG-dressing are presented in Table 4. Some recommendations regarding the way the TIG-dressing should applied have been published, [9] and [10]. Starting from these recommendations and with our own experience we have carried out the TIG-dressing on the fatigue test specimens for this project. The TIG-dressing operations have been done according to the specification illustrated in Figure 2.

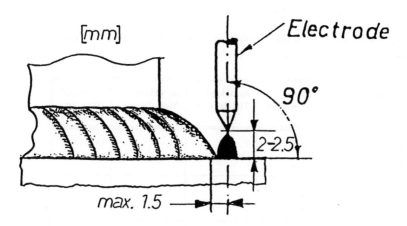

Fig. 2 Tolerance boxes for TIG-dressing procedure

It is of main concern to achieve unambiguous results regarding the application of this fatigue life improvement technique. It should be implemented in such a way that the scatter in fatigue test results should not be influenced by the the scatter in the quality of TIG-dressing. Obviously a deficiency in the application of the fatigue life improvement technique could be interpreted as an example of poor efficiency in the method itself.

Table 4 Parameters for TIG-dressing of fatigue test specimes.

Steel	Side	Pass	Electrod diameter [mm]	Ampère [A]	Volt [V]	Heat imput [kJ/mm]
DOMEX 350 DOMEX590 WELDOX 700	1,2	3,4	2.4 ARGON S 12 l/min	156	14.8	1.4

FATIGUE TESTING

The fatigue test series carried out (or on going) are presented in Table 5. The following is included: as-welded, TIG-dressed weld toe, blast-cleaned weld toe, hammer peened weld toe. The specimens in Table 5 have been tested in constant amplitude and under different spectrum loading sequences.

During the fatigue testing of improved specimens a significant influence from the parent plate surface quality on the fatigue strength has been detected. This influence has been seen both in the as-welded condition and in the TIG-dressed condition. Therefore we supplemented the original fatigue test programme with specimens there the parent plate have been blast-cleaned before welding (see Table 5). As will be seen below, this operation also influences the residual stress field at weld toe and the crack initiation site.

Table 5 Fatigue test series carried out during this investigation. The digits indicates the number of test specimens tested for the corresponding loading type.

Steel	Condition of plate surface	Load	As welded	As welded + TIG-dressed	As welded + Blast-cleaned	As welded Hammer Peened
DOMEX 350	Mill-scale unremoved	CA SP2 SP3	9 3 7	6 On going "	12	On going
DOMEX 590	Mill-scale unremoved	CA SP2 SP3	6	7 5 4	10	On going
DOMEX 590	Mill-scale blast-cleaned	CA SP2 SP3	9	On going " "	On going	On going
WELDOX 700	Mill-scale blast-cleaned	CA SP2 SP3	7	7 5	7	7

SPECTRUM DESCRIPTION

In this investigation two different spectra have been used. These are summarised in Table 6. A randomised sequence was created within each block, called SP2 and SP3, by employing a draw without replacement routine. The blocks were then repeated without reseed until fracture occurred. This gives an entirely randomised sequence to failure. In Table 6 the irregularity factor I, is defined as the number of positive mean crossings divided by the total number of cycles in one block and p is the ratio of minimun load to maximun load in a exceedance distribution according to the Swedish Building Code, BSK [11]. In Figure 3 are plotted the range pair distribution and the level crossing distributions for two utilised spectra.

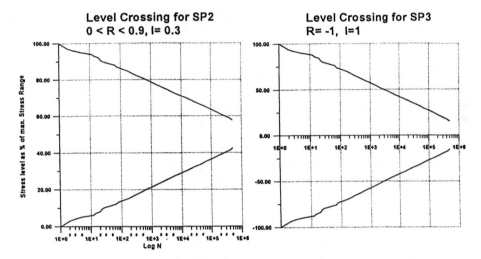

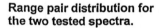

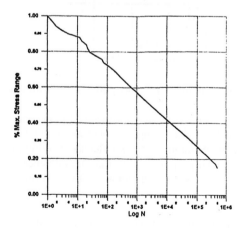

Fig. 3 Range-pair and level crossing distributions for the two spectra used.

Table 6 Spectra used in this investigation

Spectrum	Mean stress	Stress ratio	p	I	Block length
SP2	0.5 σ_{max}	0<R<0.9	1/6	0.3	5x10⁵ cycles
SP3	0	R=-1	1/6	1.0	5x10⁵ cycles

RESIDUAL STRESS MEASUREMENTS

The method used for measuring residual stresses was X-ray diffraction. The locations of the measurement points are shown in Figure 4. The radiated area was approximately 4x6mm. For this reason the first point is defined as being located 13 mm from the stiffener end. In the case of the TIG-dressed specimens it is very difficult to define the exact location of the fusion line. Therefore we have decided to measure the distance from the stiffener end in the hope we are as close as possible to the remelted weld toe. The second point was located 10 mm out from the "weld toe" in the previous measurement, and the rest of the points were located 10 and 30 mm out from the correspondent previous point. In all cases the longitudinal stress component were measured.

Three measurements on five specimens have been performed. The first measurement was carried out in the as welded condition. The second after a certain number of cycles, namely 1x10⁶. After another 5x10⁵ cycles a third measurement was carried out.

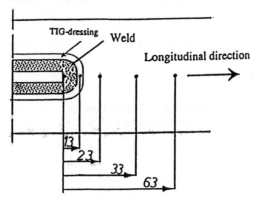

Fig. 4. Locations of residual stress measurements points in TIG-dressed specimens.

Residual stress measurements were carried out for the steel grade and different parent plate surface conditions as presented in Table 7. The change of residual stress level and shape due to the application of the TIG-dressing technique and the influence of different parent plate yield strength levels have been documented. The purpose was to find the influence of higher yield strength on the residual stress level both in the as-welded condition, in the TIG-dressed condition and in the hammer-peened condition.

The error in the stress calculation method is typically about 10 MPa. The error in the location of the x-ray beam on the sample is approximately 1 mm and the measurements give a stress gradient in the longitudinal direction about 10 MPa/mm. This gives an additional error that is superimposed on the previously mentioned one. Thus the total estimated error for the measurements is about ± 20MPa. A thorough description of residual stress measurements can be found in [12].

Table 7 Residual stress measurements carried out under this project.

Steel	Plate surface condition	Weld toe condition	Max. Load	Spectrum see Table 6	1st meas Cycl.	2nd meas Cycl.	3rd meas Cycl
DOMEX	Mill-scale	TIG-dressed	None		0		
590	Bl. Clean	TIG-dressed	425	SP2	0	1E5	5E5

RESULTS AND DISCUSSIONS

Fracture surface analysis

Fracture surface analysis has been carried out for the specimens tested in constant amplitude and spectrum loading. This analysis includes as-welded specimens as well as TIG-dressed ones. As the parent plate surface condition plays a very important roll in the type, size and distribution of weld defects, the fracture surface analysis includes two type of plate surface quality. One is the as-delivered (unremoved mill-scale) and the second is a blast-cleaned surface. This blast-cleaning operation is done at the steelwork plant in SSAB Oxelösund. For simplicity all the steel blast-cleaned surfaces have been treated at the same plant, even if the DOMEX grades tested are manufactured at the SSAB Strip Products plant at Borlänge.

As-welded specimens

For the specimens in the as-welded condition the flaws types are mostly cold laps as illustrated in Figure 5. All these defectes are comparables to the defects reported in [4] and [13]. The size and shapes vary and all the approximative measured values are presented in Figure 5.

Fatigue life improved specimens

The examined specimens have been TIG-dressed before fatigue testing. That means that weld toe region is not longer easy to localise. In the case of specimens welded and TIG-dressed (parent plate with unremoved mill-scale) the type of flaws detected are very close to those detected in the as-welded condition and with identical plate surface condition.

This situation changes radically when the parent plate has been blast-cleaned before manufacturing the specimens. The difference on the size of weld defects can be visualised from Fig. 6.

L. L. Martinez and A. F. Blom

Specimen	Type of defect	Dimensions(mm)
Domex 350-6 $\sigma_R= 250$ MPa $N= 54100$ c.	Cold lap	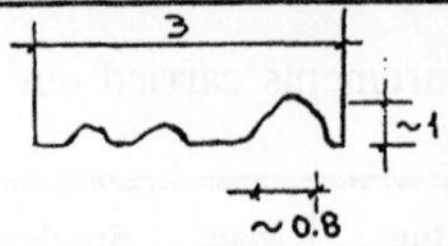
Domex 350-7 $\sigma_R= 70$ MPa $N= 3102000$ c.	Several cold laps	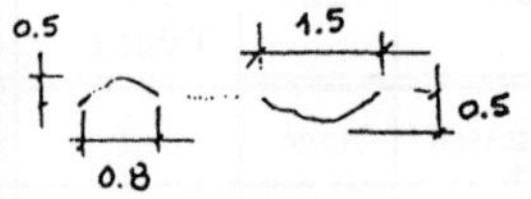
Domex 590-1 $\sigma_R= 250$ MPa $N= 54100$ c.	Cold lap	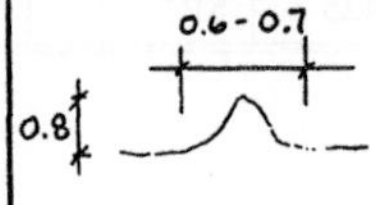
Domex 590-4 $\sigma_R= 90$ MPa $N= 1023000$ c.	Cold lap	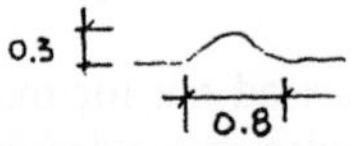
Weldox 700-2 $\sigma_R= 125$ MPa $N= 560000$ c.	Cold lap	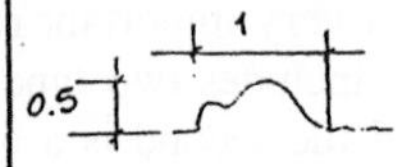
Weldox 700-8 $\sigma_R= 450$ MPa $N= 20000$ c.	Cold lap	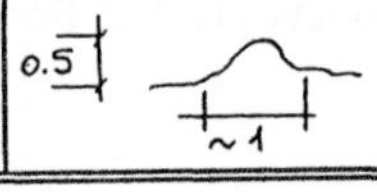

Fig. 5. Fracture surface inspection for AS WELDED specimens tested under constant amplitude and spectrum loading. Defect type and orientation.

The fracture surface analysis of the specimens in which we have applied fatigue life improvements techniques offers an unique possibility to understand the difference in the initiation life for these cracks. However, it has been documented that cold laps still exist. In the TIG-dressed specimens, some tenth of a millimeter under the remelted surface, some cold laps were still found. These have in several cases acted as crack origin sites. For these fracture surface investigations we have used a SEM6400 to clarify the influence of inclusions, plate impurities and/or if the welding consumables interact to give rise to potential flaws. Some results are illustrated in Fig. 7.

Specimen	Type of defect	Dimensions(mm)
Domex 590-2A SP2 σ_{max}= 500MPa N= 1.8E+6 c.	Deep crack at fusion line	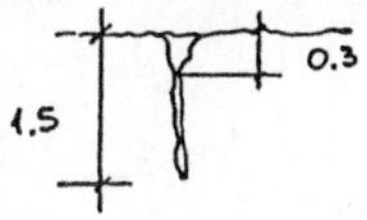 0.3 1.5
Domex 590-7A SP3 σ_{max}= 250MPa N= 9.63E+6 c.	Cold lap	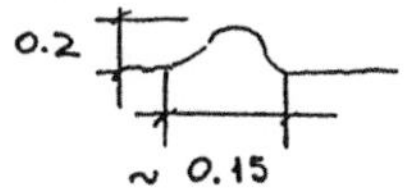 0.2 ~ 0.15
Domex 350-5 SP3 σ_{max}= 175MPa N= 4.99E+6 c.	Several cold laps	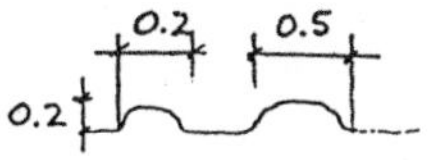 0.2 0.5 0.2
Domex 590-1 SP2 σ_{max}= 500MPa N= 1.8E+6 c.	Lack of fusion	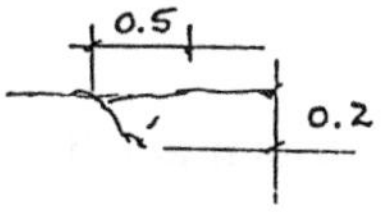 0.5 0.2

Fig. 6 Fracture surface inspection for TIG-dressed specimens tested under constant amplitude and spectrum loading. Defect type and orientation

Figure 7a Fracture surface inspection done with SEM for as-welded specimen

L. L. Martinez and A. F. Blom

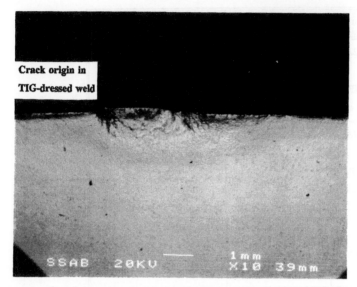

Fig. 7b Fracture surface inspection done with SEM for TIG-dressed specimen

Degree of improvement under constant amplitude

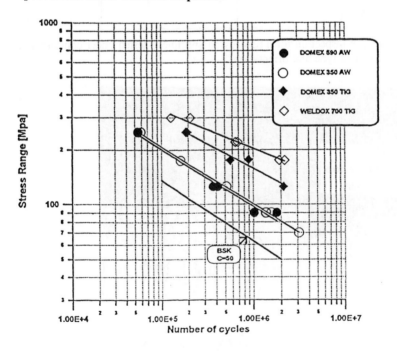

Fig. 8 Degree of improvement under constant amplitude.

The degree of improvement in fatigue life for the tested steel grades and some improvement techniques are presented in SN-diagram, Figure 8. As can be seen, the influence of material strength is remarcable when TIG-dressing is applied to steel with higher yield strength. For this comparison we assume the fatigue strength in the as-welded condition to be approx. the same for the all steel grades. As will be discussed below TIG-dressing induced some degree of compressive stresses and at the same time it improves the weld toe region. As the constant amplitude fatigue testing does not produce any noticeable relaxation of the induced compressive residual stresses, the degree of improvement can be assumed to depend on the yield level of the parent plate.

Degree of improvement under variable amplitude

The effect of fatigue life improvement techniques under spectrum loading is presented in Figure 9. The influence of material strength is not obvious from the fatigue test results so far. Nevertheless, some improvement of fatigue strength can sometimes be noted when results of the as-welded series are taken into consideration. Probably this is due to the improvement of weld toe profile. Due to relaxation of compressive stresses the effect of material strength becomes less than for constant amplitude loading. However, if the occurrence of cold laps can be minimised in the future, crack initiation life will increase and material strength may become more important.

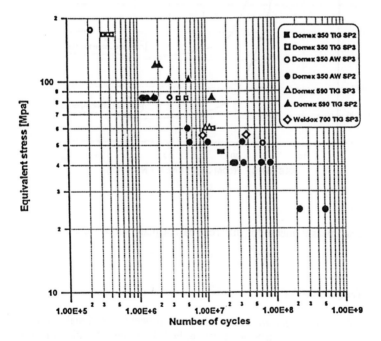

Fig. 9 Spectrum fatigue results.

Residual stress measurements previous to fatigue testing

The results of residual stress measurements are presented in Figure 10. One important result obtained from the measurements is the level of residual stresses after TIG-dressing. This can explain the high degree of improvement as well as the effect of parent plate yield strength on the TIG-dressed specimens under constant amplitude fatigue loading. Another very important result is the possible effect of blast-cleaning of the plate surface before welding. This operation as well understood induces always some level of compressive residual stresses on plate surface. This level along with the type of blast-cleaning is very dependent on the mechanical properties of the parent plate. As seen in Figure 10, the final residual stress distribution results from a combination of blast-cleaning, welding, TIG-dressing and final shotpeening (not included here). Hence, the final fatigue strength is partly determined by the initial blast-cleaning process. Nevertheless, the final effect on fatigue strength appears to be mainly determined by the type of spectrum loading.

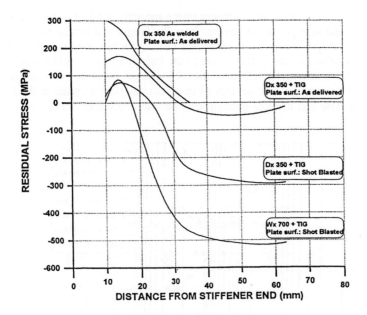

Fig. 10 Residual stress measurements for different parent plate surface conditions and improvement methods.

Relaxation under fatigue testing

The effect of the applied load spectrum in the relaxation process of residual stresses have been studied for DOMEX 590 steel. In ref[8] we have studied the same relaxation process for

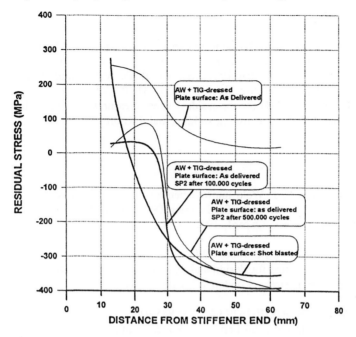

Fig. 11 Relaxation of residual stresses under spectrum fatigue loading

DOMEX 350 steel. The results are presented in Fig. 11. As can be seen in Fig. 11 most of the relaxation takes place close to weld toe where the stress concentration is largest. The main part of relaxation process occurs under the first 100.000 cycles for the tested spectrum. That is in accordance with results in [8] there 50% or more of the initial stresses are relaxed within 8% of the total life of the specimen.

CONCLUSIONS

The influence of weld flaws, in particular cold laps, on the fatigue strength of one pass weldements was shown to be great even in TIG-dressed specimens. However, futher development, both in weld quality and TIG-dressing application, is expected to substantially improve the situation.

The influence of material strength for TIG-dressed specimens is significant under constant amplitude fatigue testing.

The parent plate surface condition is important for the number and size of weld flaws as well as for the level and sign (compressive or tensile) of residual stresses.

No significant influence of material strength under spectrum testing has been documented for the tested spectra so far. However, with improved TIG-dressing and in combination with subsequent shot-peening a larger initiation life may result and material strength might become of more importance.

ACKNOWLEDGEMENTS

This work was financially suported by NI (Nordic Industrial Foundation), NUTEK (Swedish National Board for Industrial and Technical Development), SSAB, ABB and FFA. The authors are indebted to Mr. Hans Trogen (SSAB Strip Products), Mr. Bengt Wahlstenius (FFA), Mr. Tormod Dahle (ABB Corporate Research) and Mr. Sune Bodin (KTH, ILK) for performing the fatigue tests, Mrs. Annethe Billlenius (LiTH) for the residual stress measurements and Mr. Tommy Linden (SSAB Oxelösund AB) for performing welding and TIG-dressing.

REFERENCES

1.   Haagensen, P. J. Weld improvement methods for increased fatigue strength. Proc. IIW Conference "Engineering Design in Welded Constructions", Pergamon Press, Oxford, 1992, p 73-92.

2.   British Standard BS 7608: 1993
            Code of practice for design and assessment of steel structures.
            BSI, 2 Park Street, London W1A 2BS, UK.

3.   IIW document XIII-1539-94 / XV-845-94.
            Recommendations on Fatigue of Welded components.
            Prof. Dr. A. Hobbacher., Fachschule Wilhelmshaven, Fr. Paffrath-Str. 10, D-26389 Wilhelmshaven, Germany.

4.   Lopez Martinez L. and Korsgren P. "Characterization of Initial Defect Distribution and Weld Geometry in Welded Fatigue Test Specimen", Proc. of the Nordic Conference on Fatigue, Edited by A.F. Blom, EMAS Publishers, West Midlands, England, 1993.

5.   Blom, A.F., "Fatigue Strength of Welded Joints Subjected to Spectrum Loading", Jernkontoret, Stockholm, Report No. 41-02, 1987.

6.   Eide, O.I. and Berge, S., "Cumulative Damage of Longitudinal Non-load Carrying Fillet Welds", Proceedings of the Conference "Fatigue 84". Edited by C.J. Beevers, EMAS Ltd., Warley, England, 1984.

7.   Berge, S. Eide, O.I., "Residual Stress and Stress Interaction in Fatigue of Welded Joints", Residual Stress Effects in Fatigue, ASTM STP 776. American Society for Testing and Materials, 1982.

8.   Bogren J., and Lopez Martinez L., "Spectrum Fatigue Testing and Residual Stress Measurements on Non-load Carrying Fillet Welded Test Specimens". Proceedings of the Nordic Conference on Fatigue. Edited by A.F. Blom, EMAS Publishers, West Midlands, England, 1993.

9.   Nordh, B.A., Sonander C. and Sperle, J-O., "Anvisningar för TIG-behandling av svetsar för höjning av utmattningshållfastheten", SBI, Publikation 46, 1974.

10.   Lindskog, G., "TIG-behandling: Teknisk och ekonomisk bedömning av en metod för höjning av utmattningshållfastheten hos svetsförband" ,IVF-resultat 78641, November 1978.

11.   BSK, Bestämmelser för stålkonstruktioner (Regulation for steel Structure, part of the Swedish Building Code), Statens Planverk och AB Svensk Byggtjänst, 1987.

12.   Bogren, J., Lopez Martinez, L. and Brunnberg, M., "The In fluence of Various Spectrum Parameters on Fatigue Life and Residual Stress Relaxation", The Aeronautical Research Institute of Sweden, Bromma, FFA TN 1991-44.

13.   Hedegård, J., Lopez Martinez, L., Moradashkafti, N. and Trogen, H., "The Influence of Welding Parameters on the Size and Distribution of Weld Defects for Various Weld Methods". This conference.

ON MULTIAXIAL FATIGUE TESTING TIME REDUCTION

D. S. TCHANKOV
Department of Strength of Materials, Technical University of Sofia, 1156 Sofia, Bulgaria

T. GIUNTI
Processes and Materials Department, Fiat Research Centre, 10043 Orbassano (TO), Italy

ABSTRACT

This paper discusses the problem of selecting a filter level to create an edited loading history from rosette strain time histories. Parameters for obtaining the optimal filtering level that provide the biggest reduction in the testing time, but not influencing the estimated fatigue life between original and edited history (or estimated for non-edited time history) fatigue life are considered. The approach for reduction of testing time is tested with simulated and measured on a vehicle strain-histories.

KEYWORDS

Random Loading, Fatigue Testing, Multiaxial Fatigue, Multiaxial Damage Accumulation

INTRODUCTION

Requests from industry to reduce the time to market in terms of design and testing, lead to investigations on methodologies to reduce the testing time. The majority of structures in the automotive, aerospace, railroad, offshore, etc. industries are subjected to multiaxial variable amplitude service loading. Fatigue strength of structural materials is usually determined under cyclic uniaxial loading, rarely under random loading. However, for multiaxial non-proportional loading making predictions about fatigue life is very difficult and many papers are published elsewhere attempting to solve this problem [1-4]. For real components or structures the majority of the investigators carry out full-scale tests. Unfortunately, random loading biaxial and multiaxial fatigue tests are usually expensive and time consuming. Therefore the most important target is to find standard procedures to evaluate, in terms of damage, the load histories relevant on real component during the real mission and to realise more efficient load mission either for the experimental testing or in the design stage. The possibility for realising of multiaxial fatigue tests by duplication of the measured loading or strain data sequences or by reconstruction of a rainflow cycle counting matrix and a reduction of the testing time by computer editing out the data sequence is discussed in this work.

In the multiaxial state of stresses and non-proportional loading at least two independent loading histories are applied to the structure and strain gage rosette is used for fatigue data acquisition. When the measured strain sequence is edited out the major problem is that all the measured channel's information must be considered as a whole, since an individual channel data analysis will lead to misinterpretation of the information about the loading process.

BACKGROUND

For the uniaxial fatigue cases of random loading techniques for generation of loading histories and reduction of fatigue testing time are known [5-7] and they can be easily applied in proportional multiaxial loading tests. They can be summarised as: Omission of Small Non-damaging Cycles/levels; Omission of Stresses Lower than Fatigue Limit; Decreasing Time at Load; and Increasing Test Load. It should be noted that every method for reduction of fatigue testing time must be applied after previous detailed analysis and consideration of the service conditions and material and structure behaviour, for example cyclic hardening or softening. In some cases changing of material properties in the time of loading and crack growth should be also considered.

However for testing time reduction under random non-proportional multiaxial loading at the moment there are not any validated methods and algorithms. Heuler and Seeger [6] suggest that for an optimal filtering level the intrinsic fatigue level can be used. This idea is used in [7]. Thus, they ignored the loading sequence. Fash et. al. [8] investigate five irregular bending and torsion loading sequences, for using them in a test programme. There an approach of analysis by cycle counting method individually on every measured data sequences is applied. Two Markov matrices (bending and torsion loading) are presented for every loading sequences. The fulfilled fatigue damage analysis shows that a larger percent of the damages accumulate by the medium range cycles. Low range cycles, 15-20% of maximum range for smooth specimens and 25-30% of maximum range for notched specimens, give less than 5% of the total damage and can be omitted. Dressler et. al. [9] propose a new multiaxial rainflow algorithm for rainflow cycle counting on every channels simultaneously plus cycle counting on the sum and difference of every two measured data sequences. In this way the authors report that the generation of new data sequence, that is damaging equivalent to the initial one, is possible, and it is much better than an individual channel analysis and generation. But they have not presented enough test results and an approach for predicting the fatigue life, which will permit the evaluation of this new rainflow procedure. Schutz and Klatschke [10] presented results from a large research project for developing a multiaxial loading standard for car suspensions, but it is pointed out that experiments with standard load sequence are not appropriate to prove sufficient fatigue strength of components for particular applications. Widely used is the possibility to duplicate tests on structures or materials with a recorded loading history. With many problems in the past when analogue recorders have been used, for example in determining the minimal typical length of the record, the combination of different service modes and the lack of possibility to cut off some small amplitudes, today computers allow the analysis of very long loading sequences and the construction of fatigue testing programmes.

THEORY FOR DAMAGE ACCUMULATION

The choice of a theory for estimation of the fatigue life is very important. It is necessary to have a damage accumulation theory that considers non- proportional multiaxial loading. The most widely used strain and stress approaches for prediction of fatigue life under multiaxial loading can be used successfully when the multiaxial loading is proportional, but new methods, which are sensitive to the path of loading are still necessary. Such ideas have been developed by Brown and Miller[1], Socie and co-workers [2-3], Wang and Brown [4]. Socie [2] consider that a shear stress based damage parameter may to be used for calculation of fatigue damages to predict the fatigue life for materials developing shear cracks and a modified Smith-Watson-Topper parameter for materials developing tensile cracks. Also an estimation of the accumulated damages at all possible failure planes is suggested. Fatigue life is calculated for the plane experiencing the maximum damage. All necessary material parameters can be obtained from simple tests; tension-compression and torsion, or they can be calculated according to von Misses or Tresca theories. This approach has been tested and has showed good results. Wang and Brown [4] proposed a path- independent parameter that can be in use for non-proportional loading, based on Brown and Miller's critical plane theory concept [1]. Their method is more accurate and faster but there the position of the critical plane in not of great importance. These theories for damage accumulation use the Palmgren - Miner rule and they suppose that the component failure will occur when the total damage is equal to 1. Unfortunately in the service this is rarely observed and a correction is desirable. But none of the theories have been validated on real loading histories that have thousands of data points. They have only have been used on simple loading cases on tubular specimens. As result we really do not know how "accurate" they are.

EXPERIMENTS AND RESULTS

It seems that the easiest and simplest way to develop a testing sequence is to follow the techniques and ideas accepted in uniaxial fatigue. For example, the procedure given in fig. 1 can be used. When a multiaxial strain-time history is measured and stored into a computer, an editing out process starts, it have to be capable of identifying peaks and walleyes on all channels. It must allow the storage, simulation long loading histories, including randomising of different loading blocks or amplitudes and omission of non-damaging events. This algorithm gives possibility of identifying important parameters in the loading path and leads to a reduction of testing time. In general, the concepts used for uniaxial loading reductions are simply extended to multiaxial loading but some modifications are necessary. If it is accepted that the material has a loading path dependent behaviour, it must to affect

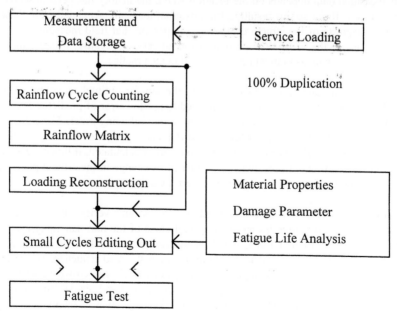

Fig.1 Algorithm for fatigue testing time reduction

programme. At any time one channel reaches a peak, values of all other channels must be stored. Also it is important that the stored data for all channels be in the same sequence according to the time [4].

As in uniaxial and multiaxial proportional loading cases, here is assumed that the small amplitudes are not damaging and they can be omitted. Therefore an editing procedure should edit out the small cycles while maintain the time sequence of the measured strains. As a basis for computing fatigue damage the algorithm and software developed by Bannantine and Socie [3] is used. It is assumed that three channels with recording strain histories data will be edited, as follows: channel 1 - ε_x, channel 2 - ε_y, channel 3 - γ_{xy}. Since the signals in each of the three channels are not proportional, a definition for multiaxial filtering level have to be chosen.

Several different possibilities were implemented in a computer code and have been tested:

-Filter level (FL) is constant value for each channel. The filter level for every channel must be set up at the beginning of the analysis.

-The filter level is a function of the maximum range of every channel's signal, which is calculated in advance. Then the three values of the filter level are different for every channel and depend only on the loading history.

-The filter level depend on the material behaviour and especially on fatigue limit. The fatigue limit is calculated according the Manson - Coffin equation for the material.

-A range limit (RL) is proposed, based on the assumption that the cycles within a range, limited by plus and minus the value of the half fatigue limit are non-damaging.

Each possibility can be used, but it should be pointed out that any random strain-time history contains information not only about the external loads but also about the dynamic response of the structure. Therefore this measured data depends on the location where the history has been observed [5,10]. It would be advantageous if the strains resulting from external loads are separated from these resulting from the dynamic response of the structure. Also it is important that field measurements contain all possible loading cases in the same time proportions as in service because, usually, measured data must be extrapolated, since measurement time is limited in the practice.

Considering that the important task is to find out a criterion for optimal reduction of the loading sequence and the testing time, as a parameter is decided to use the error of predicted fatigue life. The problem for an optimal testing time reduction is solved in the following order: The editing procedure starts with an equal increment of the filter level; the difference /change/ of the fatigue life prediction error for current filter level and the error for the previous one is calculated; if this difference exceed a user defined value, the editing out procedure is terminated. A change in the critical plane direction is not allowable. The limit for the fatigue life prediction error can be set up, usually 5%.

Numerical analysis on a data sequence is carried out to test and verify the proposed ideas and possibilities for reducing the fatigue testing time by omissions of the non-damaging cycles. A strain history with irregularity coefficient i=0.3, which is the common one, is simulated using rainflow reconstruction software and a 32-level matrix [11] for Gausian standard sequence. The received strain data is shown in fig.2. In this paper the approach developed by Socie and co-workers [2-3] is applied since it uses material parameters derived by simple tests in uniaxial and biaxial (torsion) loading and it allows the location of the critical plane to be identified. Damage analysis using material properties of three steels is performed. The mechanical properties are listed in Table 1. For example, material

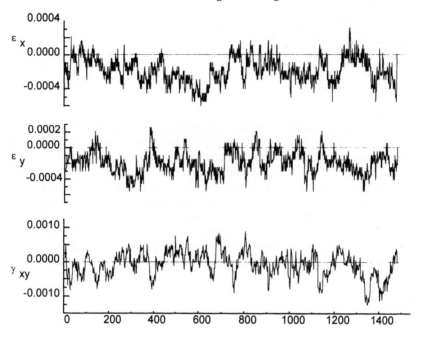

Fig. 2. Simulated Gaussian random strain history, i=0.3

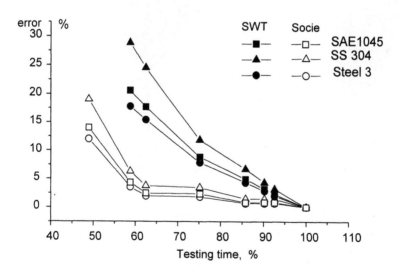

Fig.3. Estimated fatigue life prediction error vs length of the testing time

the editing SAE 1045 is chosen as it develops a shear type (Mode III) cracks. Some results representing the estimated error of predicted fatigue life versus the normalised number of cycles in one loading block are shown in fig. 3, where error is estimated as

$$error\% = \frac{N_{pr}^{edited} - N_{pr}^{nonedited}}{N_{pr}^{edited}} \cdot 100\%,$$

and testing time is estimated as

$$Testing\ time\% = \frac{T^{edited}}{T^{nonedited}} \cdot 100\%.$$

Relationship between testing time and filter level depends on the loading history and it is not shown here. It is obviously the difference between the two fatigue life prediction methods. The tensile model (SWT) gives a higher estimation and therefore a lower reduction of the testing time.

It is clear that for small filter levels both damage criteria give approximately the same fatigue life predictions. But it can be seen too, that a reduction of the testing programme with less than a specific number of cycles per block will cause a rapid increase of the error of predicted fatigue life for this material. Such reduction of the testing programme must be avoided. This fact is used as criterion to stop cycle edition out of the loading block; therefore the value of the filtering level depends on a) the loading/strain history, b) the material behaviour, c) assumed theory for damage accumulation.

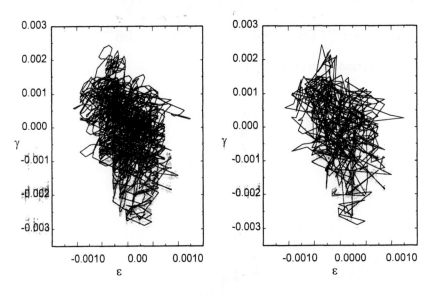

Fig.4. ε - γ plot for   a) non-edited,   b) edited out history

In fig. 4 the original and edited out paths of $\varepsilon - \gamma$ are shown. Here it can be seen that the omitted small cycles do not influence the loading path and can be neglected. In support of such judgement are the results from fatigue life predictions. Figure 5 shows the damage distribution for different planes and different filter levels. They show a critical plane has remain identical, for example $\theta= 90°$ and $\varphi=130°$ for tensile model, and this means that the loading path is not affected. If any changes in critical plane direction are observed, it means that no further edition of this loading history is possible.

Above described algorithm is applied for reduction of the testing time when a fork-lift truck was tested [12] on a 4-channel road simulator. The movement on three different road surfaces: asphalt,

Table 1. Material Data

Material Parameter	SAE 1045 [3]	SS 304 [3]	En15R [16]	Steel 3
E GPa	206.9	183	205	214
K' MPa	1255.2	1660	1336	988
n'	0.2	0.287	0.17	0.207
σ_f' MPa	948.4	1000	1114	837
b'	-0.092	-0.114	-0.097	-0.100
ε_f'	0.26	0.171	0.259	0.557
c'	-0.445	-0.402	-0.515	-0.518
τ_f' MPa	470	709	870.3	494
γ_f'	0.41	0.413	0.518	0.83

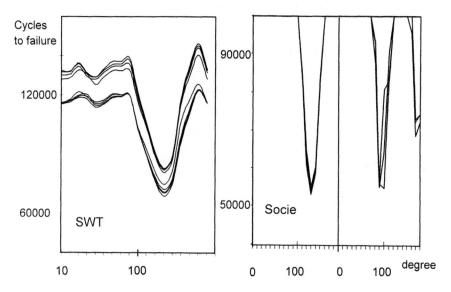

Fig. 5. Damage distribution plot according Smith-Watson-Topper (SWT) and Socie damage parameters.

pavement I and pavement II was simulated. To evaluate the non-damaging cycles in a hot spot of the fork-lift truck were measured and recorded three strain histories from a strain gage rosette for these road surfaces. The strain histories are shown in fig. 6. The strain data was used for subsequent fatigue testing. Damages were calculated for the used material steel 3 and SAE1045. Since a slightly different critical plane is predicted for asphalt, the order of the loading block sequence was examined. The results show that the order can be important and the service conditions should be followed strictly. If there is not enough information for the service conditions, for example on the design stage, a "randomised" loading block can be used. Fig. 7 shows the relationship between error of the predicted fatigue life and the normalised testing time for asphalt, pavement I, pavement II and a joint randomised sequence.

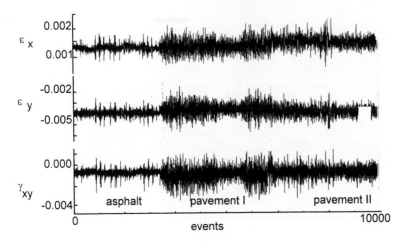

Fig. 6. Fragment of measured strain history on a fork-lift truck

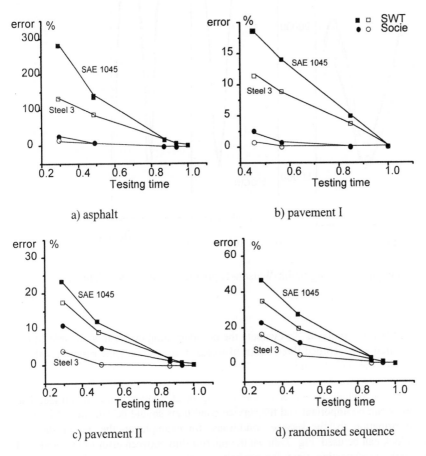

a) asphalt

b) pavement I

c) pavement II

d) randomised sequence

Fig. 7. Estimated fatigue life prediction error versus the time of testing

CONCLUSION

An important problem is considered: reduction of the time for fatigue testing of materials and components. A new criterion for optimal filter level value is proposed. It allows to be implemented easily and decision for the value of the filter level would be made after few fatigue life calculations using a rule for damage accumulation under multiaxial state of stresses. Editing out of strain time histories under multiaxial stresses conditions and omission of small cycles have been proposed. The results are encouraging, this technique permits reducing of the necessary time by a factor from 2 to 8. Different possibilities about the filter level have been suggested and tested on a strain sequence. Further reduction of the chosen testing programme can be done by proportional increasing of the load.

This interdisciplinary field will be ever more important and it will be supported from technical people able to understand the integration with experimental experience and the new developed methodologies.

ACKNOWLEDGEMENTS

Part of this work was carried out under the ECC "Go-West" fellowship No CIPACT920352/7591. The authors would like to thank Dr. A. Blarasin for his comments and Prof. D.F. Socie for the permission to use his software.

REFERENCES

1. Brown M.W., Miller K.J., (1973) *Proc. Instn. Mech. Engrs.*, No 187, pp. 745-755.
2. Socie D.F., (1987) *Trans. ASME, J. of. Eng. Mater. and Tech.*, **109**, pp 293-298.
3. Bannantine J.A., Socie D.F., (1992) In: *Advances in fatigue lifetime predictive techniques*, M.Mitchel and R.Landgraf (Eds), Philadelphia, ASTM STP 1122, pp249-275.
4. Wang C.H., Brown M.W., (1993) *Fatigue Fract. Engng. Mater.Struct.* 16, No.12, pp. 1285-1298.
5. Chankov D.S., Final Report, CRF, Orbassano, Oct., 1993, 35 p.
6. Heuler P., Seeger T., (1986) *Int. J. Fatigue*, **8**, No 4, pp. 225-230.
7. DuQuesnay D.L., Pompetzki M.A., Topper T.H., (1993) SAE paper 930400, SAE, Warrendale.
8. Fash J.W., Conle F.A., Minter G.L., (1989). In: *Multiaxial Fatigue: Analysis and Experiments*, AE - 14, D. Socie and Leese (Eds), SAE, Warrendale, Pasadena, pp 33-60.
9. Dressler K., Carmine R., Krueger W., (1992). In: *Low Cycle Fatigue and Elasto-Plastic Behaviour of Materials*, Ed. K.T.Rie, Berlin, pp 522-530.
10. Schutz W., Klatschke (1992). H., In: *Proc. 3-rd Int. Conf. Innovation and Reliability in Automotive Design and Testing*, Florence, Italy, pp. 1231-1250.
11. Olagnon M., (1994) *Int. J. Fatigue*, **16**, July, pp. 304-313.
12. Vesselinov K.V., et.al., Report TH32, TU Sofia, 1994, 60p. (In Bulgarian).

STANDARD FATIGUE TESTS FOR COMPONENTS

L P POOK
University College London
Department of Mechanical Engineering, Torrington Place, London WC1E 7JE, UK

ABSTRACT

A recent review of current British Standards showed that there are at least 52 documents which contain clauses requiring acceptance fatigue testing of components. Most of these clauses appear in product standards. Six acceptance fatigue tests are outlined and discussed. The examples chosen include components from both industrial and consumer items, and are: high strength steel bolts, hydraulic filter housings, light alloy car wheels, electrical switches, cookware handles, and heels of ladies' shoes. It is concluded that the tests are a compromise between accurate representation of service conditions, and the need to keep the cost and duration of testing to a minimum. In particular, constant amplitude loading is generally employed.

KEYWORDS

Components, Fatigue tests, Standards, Bolts, Hydraulic filters, Car wheels, Electrical switches, Cookware, Shoes.

INTRODUCTION

There are three main approaches to the fatigue asessment of components[4]; an analytic aproach, use of a standard design procedure, and service loading testing. In practice, some combination of these is normally used. Whatever approach is used, allowance must be made for the inevitable scatter in the fatigue lives of specimens and components, and factors such as uncertainties in service loads and stress analysis.

An analytic approach makes use of information on service loads and material properties, and applied mechanics. It requires expert knowledge. The fatigue crack initiation and propagation phases must be considered separately, although in practice one or the other usually predominates. Typically, elaborate calculations are necessary, and the approach may fail because not all the detailed information required is available. From the designer's viewpoint standard design procedures are the most satisfactory. These range from informally established 'good practice' in a particular design office to elaborate codes, often imposed by regulatory authorities. Standard design methods have the advantage that less expert knowledge is required, and they can conveniently be incorporated in software packages.

Modern test equipment permits the application of virtually any desired load history, so service loading testing can be used for fatigue asessment when an analytic approach and standard design procedures fail[4]. Fatigue testing of prototypes, or production samples, before components are put into service, is widely used as an acceptance test for critical components[3]. It is sometimes a requirement of regulatory authorities, and has the advantage that its basis is easily understood by laymen. Another advantage is that weak points in a design can be identified and rectified at the prototype stage.

A striking feature of analysis of documents issued by the British Standards Institution[5, 6]. is the large number which include clauses requiring acceptance fatigue testing of components. A review (February 1995) of current British Standards showed that there are at least 52 documents which contain such clauses. Most of them are product standards. The documents located are listed in the Appendix.

Six acceptance fatigue tests are outlined and discussed. The examples chosen include components from both industrial and consumer items, and are: high strength steel bolts, hydraulic filter housings, light alloy car wheels, electrical switches, cookware handles, and heels of ladies' shoes.

EXAMPLES

High Strength Steel Bolts

The example chosen is: BS A 241: 1973. General requirements for steel protruding-head bolts of tensile strength 1250 MPa (180 000 lbf/in^2) or greater. This gives requirements for high strength steel bolts for aerospace use. The acceptance fatigue test specified is a tension-tension test. A suitable test rig is shown schematically in Fig. 1. The maximum load is the stress area of the bolt × 650 MPa for bolts with a tensile strength of 1250-1380 MPa, and the stress area × 900 MPa for bolts with a tensile strength of 1800-1930 MPa. The minimum load is 10 per cent of the maximum load. The test frequency is 500-12,500 cycles per minute, and tests are continued to 130,000 cycles, or to failure, whichever occurs first. The number of bolts in the samples tested depends on the size of the batch being evaluated. For example, for a batch size of 501-1300 the first sample size is 7, and the second sample size is 14.

Test results are evaluated as follows. For the first sample the batch is accepted if the geometric mean life of the sample is > 65,000 cycles, and if the minimum individual bolt life is > 45,000 cycles. It is rejected if the geometric mean life is ≤ 65,000 cycles, or if two or more bolts have lives of ≤ 32,500 cycles. A second sample is taken only if the batch is neither accepted nor rejected on the basis of the test results from the first sample. For the second sample the batch is accepted if the geometric mean life of the combined samples is > 65,000 cycles, and if the minimum individual bolt life in the second sample is > 32,500 cycles. It is rejected if the geometric mean life of the combined samples is ≤ 65,000 cycles, or if the minimum individual bolt life in the second sample is ≤ 32,500 cycles. In all cases the geometric mean life is calculated from the actual cycles to failure, or 130,000 cycles, whichever is the lesser.

Comment. A bolt is a ubiquitous engineering component. It is not surprising that a long established standard for high quality high strength steel bolts includes an acceptance fatigue test. Sophisticated statistical criteria are used to determine the acceptability of a particular batch of bolts. The constant amplitude loading specified is obviously not intended to represent any particular service loading. The maximum test duration is 4½ hours.

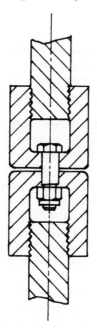

Fig. 1. Fatigue test rig for a bolt.

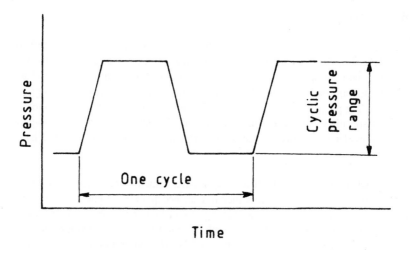

Fig. 2. Pressure cycle for fatigue test on a hydraulic filter housing.

Hydraulic Filter Housings

The example chosen is: BS 7268: 1990. Method for determination of fatigue pressure rating of metal pressure containing envelopes in hydraulic fluid power systems. This was prepared as the result of research, during which it became apparent that both the frequency and the waveform have a pronounced effect on the internal pressure fatigue life of metal hydraulic fluid power components.

The standard includes an acceptance test for determination of the fatigue pressure rating of various components including hydraulic filter housings. The fatigue pressure rating is the specified number of housings which have been tested to a specified number of cycles at a specified cyclic pressure range at a specified cycling rate, for example '5 at 2×10^6 at 315 bar at 1.2 Hz'. At least three housings must be tested for type approval. The pressure cycle has the waveform shown in Fig. 2. The minimum pressure must not exceed 10 per cent of the cyclic pressure range or 16 bar. The maximum permissible viscosity of the pressurizing medium is 60 centistokes at the test temperature, and the pressure cycling rate must not exceed 1.5 Hz. Providing that failure does not occur earlier the test is continued for 10^6 cycles (unless otherwise specified). Failure is defined as the structural fracture of a housing or the production of any fatigue crack produced by pressure cycling, as found by normal crack detection techniques.

Comment. A hydraulic filter housing is an example is of a safety related industrial component. The standard, which also covers other hydraulic components was developed from an earlier standard issued by the British Fluid Power Association[7, 8]. The constant amplitude loading specified does not represent any particular service loading. The standard uses fatigue tests to define a fatigue pressure rating rather than precise acceptance criteria. The maximum test duration is 8 days, assuming that 10^6 cycles are applied. This is the default number of cycles recommended in the standard. A trapezoidal waveform is specified for historical reasons. It cannot readily be changed because internal pressure fatigue behaviour is waveform dependent[8].

Light Alloy Car Wheels

The example chosen is: BS AU 50, Part 2, Section 5b: 1976. Road wheels manufactured wholly or partly of cast light alloy for passenger cars. This includes a radial fatigue aceptance test. For this test the wheel is fitted with an appropriate tyre, inflated to not more than 455 kPa, and a constant radial force, which rotates around the wheel, is applied. This force is 2.25 × the design maximum static loading. The sampling procedure is not specified. There is no description of a suitable test rig, and no test frequency is specified. The wheel must withstand 500,000 cycles, and at completion of the test there must be no evidence of fatigue cracks anywhere on the wheel, as indicated by a dye penetrant test.

Comment. A car wheel is an example of a safety related vehicle component. It is not surprising that a long established standard for car wheels includes an acceptance fatigue test. A constant amplitude loading is only representative of a vehicle running at constant speed on a straight and level road. The number of cycles specified is small compared with the number of service cycles, but as compensation a high load is specified. Assuming a test frequency of 1 Hz, the maximum test duration is 6 days.

Electrical Switches

The example chosen is: BS 3676. Switches for household and similar fixed electrical installations. Part 1: 1989. Specification for general requirements. This includes a normal operation acceptance test. In this test switches make and break a resistive load equal to their rated current, at their rated voltage, in a substantially non-inductive a. c. circuit. Three samples are tested for type approval. Switches are operated for a specified number of operations (30,000 for a rated current ≤ 16 A), at a rate of 10-12 operations per minute with the 'on' and 'off' periods being approximately equal. A suitable test test rig for a tumbler switch is shown schematically in Fig. 3. The acceptance criterion is that a switch must remain operational, and mechanically and electrically sound, throughout a test.

Comment. An electrical switch is an example of an electrical component with mechanical parts which might fail in fatigue due to repeated operation. The test specified is defined in terms of a number of normal operations. It is effectively a constant amplitude fatigue test on the mechanical components of the switch. The number of operations specified appears to be realistic for, for example, a domestic light switch. The maximum test duration is 50 hours.

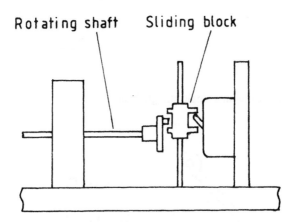

Fig 3. Normal operation test rig for a tumbler switch.

Cookware Handles

The example chosen is: BS 6743: 1987. Performance of handles and handle assemblies attached to cookware. This was prepared in response to accident statistics which demonstrate that serious accidents can occur as the result of the premature failure of handles of domestic or commercial cookware. It sets levels of performance for cookware for use on top of a stove, cooker or hob by the accelerated simulation of hazards experienced in normal use. The acceptance fatigue test specified involves continuously raising and lowering a loaded item of cookware from a level surface once per minute by means of its handle. The load is 40 mm diameter metal spheres placed in the cookware, whose mass is equivalent to 2.25 times the mass of water at the gross capacity. A suitable test rig for a pan with a single straight handle is shown schematically in Fig. 4. Similar equipment is used for other types of handle. Two samples are tested for type approval. Each handle is required to withstand 1500 raising and lowering cycles without permanent distortion or loosening of the handle or its fixing system.

Comment. A cookware handle is an example of a safety related item in both commercial and domestic use. A constant amplitude test is specified. The number of cycles applied is quite small compared with the likely number of service cycles, but in compensation a high load is specified. The way the load is applied is a realistic representation of service loading. A square waveform is used for experimental convenience. The maximum test duration is 25 hours.

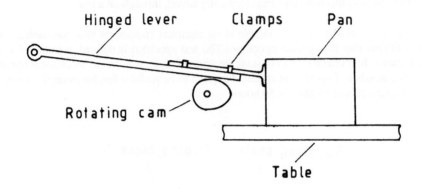

Fig 4. Fatigue test rig for a pan with a single straight handle.

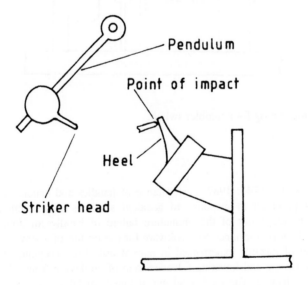

Fig. 5. Fatigue test rig for a heel.

Heels of Ladies' Shoes

The example chosen is: BS 5131, Part 4, Section 4.9: 1991. Fatigue resistance of heels of ladies' shoes. This describes a method for determining the ability of heels of ladies' shoes to withstand the repeated small impacts imposed by normal walking. It is particularly useful for injection moulded plastic heels which incorporate a steel dowel reinforcement. The acceptance fatigue test specified involves subjecting the heel to impacts 6 mm from the heel tip, delivered by a pendulum at the rate of one blow per second. The pendulum dimensions and fall height are specified, and each impact has an energy of 0.78 J. A suitable test rig is shown schematically in Fig. 5. Three samples are tested for type approval. The number of impacts to failure is noted, and a test is discontinued after 20,000 impacts. Any damage which develops is described.

Comment A ladies' shoe heel is an example of a safety related consumer item. A constant amplitude test is specified. The number of cycles applied is small compared with the likely number of service cycles, but in compensation a severe impact is specified. The impact velocity is not specified, but is controlled through other parameters The way the load is applied is a realistic representation of service loading. The maximum test duration is 5½ hours.

DISCUSSION

Fatigue failures of components are always a nuisance, and can be expensive and dangerous. It is therefore not surprising that there are a large number of standards, written by various organisations, which contain clauses on acceptance fatigue testing of components[5, 6]. Discussion is restricted to documents issued by the British Standards Institution. It is appreciated that documents issued by other standards writing organisation, for example the (American) Society of Automotive Engineers, are also in use in the UK.

It is well known[2] that constant amplitude tests, as compared with realistic variable amplitude tests, can give a misleading impression of the relative fatigue strengths of components, especially when the detail design or the material is changed. Nevertheless, constant amplitude loading is specified in all the acceptance fatigue tests for components contained in documents listed in Appendix 1. This apparent contradiction is resolved when commercial considerations are taken into account. From a commercial viewpoint, an acceptance fatigue test needs to be simple, quick, and cheap. The risk of a fatigue failure in service may be kept low by making a simple test relatively severe. In many situations this will minimise the overall product cost, despite possible overdesign. The 6 tests outlined were chosen to be representative of those described in British Standards. In all of them the maximum test duration is relatively short, which minimises both the duration and cost of testing. It is only in industries where minimum weight design is important, such as the aircraft industry[1], that expensive and time consuming variable amplitude tests are economically justified.

As is well known[2], for many metallic materials the results of a fatigue test are largely independent of the test frequency and waveform. Sinusoidal loading is conventional in most constant amplitude fatigue testing of specimens and components and, although not explicitly specified, is used for the acceptance tests on bolts and car wheels. A square waveform is used for the cookware test for experimental convenience. The electrical switch test is an operational test. so the waveform is not specified explicitly. Internal pressure fatigue behaviour is both frequency and waveform dependent[8]. This is a result of the time required for the pressurizing fluid to flow into and out of any fatigue cracks that may form. A trapezoidal waveform is conventionally used for internal pressure fatigue tests for historical reasons[8], and is therefore specified for the filter housings. Plastics are often strain rate sensitive[9], an impact fatigue test is therefore specified for the heels of ladies' shoes as a realistic representation of service loading.

CONCLUSIONS

There are a large number of standards, written by various organisations, which contain clauses on acceptance fatigue testing of components. Acceptance fatigue tests are a compromise between accurate representation of service conditions, and the need to keep the cost and duration of testing to a minimum. In particular, constant amplitude loading is generally specified in British Standards. Constant amplitude tests can give a misleading impression of the relative fatigue strengths of components. However, the risk of a fatigue failure in service may be kept low by making a constant amplitude test relatively severe. In many situations this will minimise the overall product cost, despite possible overdesign. The existence of accceptance fatigue testing standards has undoubtedly reduced the number of fatigue failures in service. For example, 35 years ago fatigue failures in motor vehicles were commonplace, but this is no longer the case.

REFERENCES

1. Beukers, A., de Jonge, Th., Sinke, J., Vlot, A. and Vogelesang, L. B. (Eds.). (1992). *Fatigue of Aircraft Materials*. Delft University press, Delft.

2. Frost, N. E., Marsh, K. J. and Pook, L. P. (1974). *Metal Fatigue*. Clarendon Press, Oxford.

3. Marsh, K.J. (Ed.). (1989). *Full-Scale Fatigue Testing of Components and Structures*. Butterworth Scientific Ltd, Guildford.

4. Pook, L .P. (1983). The *Role of Crack Growth in Metal Fatigue*. Metals Society, London.

5. Pook, L, P. (1989). *Current Fatigue Standards 1989*. Engineering Integrity Society, Daventry.

6. Pook, L. P. (1989). In: *Proceedings of the National Fatigue Symposium*, Solin, J. (Ed.). Technical Research Centre of Finland, Espoo, pp. 16-22.

7. Pook, L. P. (1989). In: *Proceedings of the National Fatigue Symposium*, Solin, J. (Ed.). Technical Research Centre of Finland, Espoo, pp. 104-110.

8. Pook, L. P. and Short, A. M. (1988). In: *Fluid power 8*. Heron, R. (Ed.). Elsevier Applied Science Publishers, London, pp. 323-338.

9. Williams, J. G. (1984). *Fracture Mechanics of Polymers*. Ellis Horwood, Chichester.

APPENDIX BRITISH STANDARDS

Documents, issued by the British Standards Institution, which contain clauses on acceptance fatigue testing are listed below. The list was updated in February 1995.

Automative Components

BS AU 24a: 1989. Towing connections for trailers up to 5000 kg gross mass.

BS AU 50. Tyres and wheels. Part 2. Wheels and rims. Section 5b: 1976. Road wheels manufactured wholly or partly of cast light alloy for passenger cars.

BS AU 50. Tyres and wheels. Part 2. Wheels and rims. Section 6: 1983. Road wheels manufactured wholly or partly of cast light alloy for mopeds and motor cycles.

BS AU 114b: 1979. Strength requirements of towing brackets and coupling balls for caravans and light trailers.

BS AU 235. Fifth wheel couplings for commercial vehicles. Part 1: 1989. Test conditions and strength requirements.

Bolts

BS A 101: 1969. General requirements for titanium bolts.

BS A 241: 1973. General requirements for steel protruding-head bolts of tensile strength 1250 MPa (180 000 lbf/in^2) or greater.

BS A 274: 1981. Procurement of alloy steel bolts, metric, with a minmum tensile strength of 1100 MPa.

Electrical Switches

BS 3676. Switches for household and similar fixed electrical installations. Part 1: 1989. Specification for general requirements.

BS EN 2495: 1991. Single-pole circuit breakers temperature compensated rated currents up to 25 A.

BS EN 2592: 1991. Three-pole circuit breakers temperature compensated rated currents up to 25 A.

Footwear Parts

BS 5131. Methods of test for footwear and footwear materials. Part 4. Other components. Section 4.9: 1991. Fatigue resistance of heels of ladies' shoes.

BS 5131. Methods of test for footwear and footwear materials. Part 5. Tests of complete footwear. Section 5.7: 1978. Fatigue tests for rigid units and shoe bottoms.

Hydraulic Components

BS 3832: 1991. Wire reinforced rubber hoses and hose assemblies for hydraulic installations.

BS 4368. Carbon and stainless steel compression couplings for tubes. Part 4: 1984. Type test requirements.

BS 4552. Fuel filters, strainers and sedimentors for compression ignition engines. Part 1: 1979. Methods of test.

BS 5173. Methods of test for rubber and plastics hoses and hose assemblies. Section 102.5: 1985. Pressure impulse test for high pressure hydraulic hoses.

BS 5173. Methods of test for rubber and plastics hoses and hose assemblies. Section 102.7: 1988. Pressure impulse test with flexing for high pressure hydraulic hoses (half omega configuration).

BS 5173. Methods of test for rubber and plastics hoses and hose assemblies. Section 102.8: 1987. Pressure impulse test for rigid helix reinforced thermoplastics hoses.

BS 6275. Hydraulic fluid power filter elements. Part 2: 1984. Methods of test to verify structural integrity.

BS 6501. Flexible metallic hose assemblies. Part 1: 1991. Corrugated hose assemblies.

BS 6784: 1986. Rubber hoses and hose assemblies for automobile power steering systems.

BS 7268: 1990. Method for determination of fatigue pressure rating of metal pressure containing envelopes in hydraulic fluid power systems.

BS EN 26802: 1993. Rubber and plastics hose assemblies - Wire reinforced - Hydraulic impulse test with flexing.

BS EN 28029: 1993. Rubber and/or plastics hose assemblies for airless paint spraying.

BS M 55: 1984. Rotary flexure testing of hydraulic tubing joints and fittings for aerospace use.

BS 2M 60: 1993. Separable tube fittings 24° cone, for fluid systems.

BS M 65: 1987. Pressure compensated variable delivery hydraulic pumps.

Miscellaneous Components

BS 4831: 1985. Shackle type connector units for chain conveyors for mining.

BS 5950. Structural use of steel work in building. Part 4: 1994. Code of practice for design of composite slabs with profiled steel sheeting.

BS 6743: 1987. Performance of handles and handle assemblies attached to cookware.

BS 6888: 1988. Methods for calibration of bonded electrical resistance strain gauges.

DD 171: 1987. Guide to specifying performance requirements for hinged or pivoted doors (including test methods).

Orthopaedics Prostheses

BS 7251. Orthopaedic joint prostheses. Part 5: 1990. Method for determination of endurance properties of stemmed femoral components of hip joint prostheses with application of torsion.

BS 7251. Orthopaedic joint prostheses. Part 6: 1990. Method for determination of endurance properties of stemmed femoral components of hip joint prostheses without application of torsion.

BS 7251. Orthopaedic joint prostheses. Part 10: 1992. Method of determination of endurance properties of the head and neck region of stemmed femoral components of hip joint prostheses.

BS 7251. Orthopaedic joint prostheses. Part 11: 1993. Endurance of stemmed femoral components without application of torsion.

Passenger Lift and Conveyor Components

BS 5655. Lifts and Service Lifts. Part 1: 1979. Safety rules for the construction and installation of electric lifts. (Remains current.)

BS 5655. Lifts and Service Lifts. Part 1: 1986. Safety rules for the construction and installation of electric lifts.

BS 5655. Lifts and Service Lifts. Part 2: 1988. Safety rules for the construction and installation of hydraulic lifts.

BS 5656: 1983. Safety rules for the construction and installation of escalators and passenger conveyors.

Small Craft Steering System Components

BS EN 28848: 1993. Small craft - Remote steering systems.

BS EN 29775: 1993. Small craft - Remote steering systems for single outboard motors of 15 kW to 40 kW power.

Wires, Wire Ropes, Rods, Etc

BS 302. Stranded steel wire ropes. Part 1: 1987. General requirements.

BS 2763: 1982. Round carbon steel wire for wire ropes.

BS 4447: 1973. The performance of prestressing anchorages for post-tensioned construction.

BS 4449: 1988. Carbon steel bars for the reinforcement of concrete.

BS 5281: 1975. Ferrule secured eye terminations for wire ropes.

BS 5896: 1980. High tensile steel wire and strand for the prestressing of concrete.

BS 7035: 1989. Code of practice for socketing of stranded steel wire ropes.

BS 7166: 1989. Wedge and socket anchorages for wire ropes.

BS EN 2569: 1991. Control cable fittings and turnbarrels.

A NEW METHOD FOR FATIGUE LIFE IMPROVEMENT OF SCREWS

Shin-ichi Nishida*, Chikayuki Urashima** and Hidetoshi Tamasaki***

* Faculty of Science & Engineering, Saga University, Saga, 840, Japan
** Yawata R&D Laboratory, Nippon Steel Corp., Kitakyushu, 804.
*** Production & Technical Dept., Nippon Steel Bolten, Yukihashi, 824.

ABSTRACT

Motivated by the failure of several giant bolts, a new method to improve the fatigue strength of bolts has been developed. The new method is named CD bolt and stands for "Critical Design for Fracture". Improved thread profile greatly reduces the peak stresses. Fatigue limit testing has confirmed that the new profile approximately doubles the fatigue strength of bolts as compared to more traditional profiles. Improved strength is for both the pre-stressed and non pre-stressed conditions.

KEY WORDS

Bolt, Nut, Screw, Fatigue life improvement, Critical design, CD bolt

INTRODUCTION

Abrupt failure due to fatigue ocurred in two giant bolts, called tie rods, in a rolling mill. Outside thread diameter was 478 mm and the bolt length was 13,975 mm. Motivated by this incident, the authors studied numerous failed bolts. It is not an understatement that the smooth operation of machines at industrial plants is often dependent on a single bolt and that bolt failure is most often attributable to fatigue.

Effects of thread profile and other factors on fatigue strength have been investigated for bolts showing low strength. Failure can be attributed to uneven load sharing among the threads, concentration of tensile and bending stresses, and localized loading. A new method for improvement of fatigue strength of bolts has been developed. The above method is named CD bolt and stands for "Critical Design for Fracture" and relys on a uniquely optimized thread profile.

EXPERIMENTS

Chemical composition of the materials used for the test are listed in Table 1. Materials used for the nut and bolt are mainly SCM440 and SNCM630. In addition, S20C was used to study the effect of partial damage of the nut on the fatigue strength. All specimens were taken at the depth of 200 mm from the surface of the bar in such a way that the longitudinal direction of the bar became the central axis of the specimen. Threads were introduced by turning. Mechanical properties of the materials are listed in Table 2. Fatigue tests have been conducted to determine the effect of four items: thread type, root radius, nut and bolt material, and prestressing.

Table 1. Chemical composition of material used (wt%)

steel	C	Si	Mn	P	S	Ni	Cr	Mo
SCM440 (φ 455)[*]	0.41	0.35	0.73	0.0013	0.020	0.08	1.02	0.21
SNC630 (φ 470)	0.29	0.25	0.44	0.009	0.006	2.97	2.98	0.59
S20C (φ 40)	0.19	0.01	0.41	0.008	0.005	-	-	-

[*]numbers in parenthesis denote the size of the material

Table 2. Mechanical properties

steel	yield strength (MPa)	tensile strength (MPa)	elongation (%)	reduction in area (%)	impact value (J)	
					$_vE20°C$	$_uE20°C$
SCM440	582	791	21.0	-	26.5	-
SNC630	892	1009	22.0	61.4	-	100.9
S20C	> 245[*]	>402[*]	>28[*]	(Hv10)203	-	-

[*] specified value

Effect of Type of Thread

The effects of triangular thread, trapezoidal thread, positive buttress thread and negative buttress thread on fatigue strength of bolts have been studied. Profiles of these thread types are shown in Fig. 1. The outside diameter and root radius differred slightly for some of the threads but the root diameter of all the threads was a constant 25 mm.

Effect of Root Radius

Three root radii r -0.30, 0.50 and 0.70 mm - were selected for the triangular thread. The outline profile of thread with different root radii are shown in Fig. 2.

Effect of Nut and Bolt Materials

In most tests, both nut and bolt were made of SCM440 steel. A study was also carried out using a steel with higher strength (SNCM630) for the bolt and a material with lower strength (S20C) for the nut.

Effect of Prestressing

Prestressing is an effective means of improving fatigue resistance, but it is not widely adopted on site. Prior to fatigue testing, an axial tensile stress was statically applied to the nut and bolt. Prestresses of 421 and 363 MPa were selected for the combination of nut and bolt made of SNCM630 and for the combination of bolt made of SNCM630 and nut made of S20C, respectively. All stresses were presented by the nominal stresses at the root diameter (25mm). All specimens were subjected to partially tensile pulsating fatigue test with a mean stress σ_m. For the test, a servo-type fatigue test machine (±392kN) was used. The frequency was 500 cycles/min. The S-N curves were obtained for all specimens. The fatigue strength of the specimens at $2×10^6$ cycles was compared.

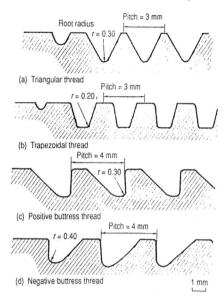

Fig. 1. Outline of thread profile of various types of thread.

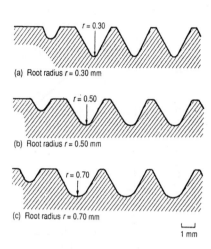

Fig. 2. Outline of thread profile with different root radii.

RESULTS AND DISCUSSION

Factors Related to the Fatigue Strength of Bolts

Figure 3 shows the effect of different kinds of screws on fatigue strength. In the case of the triangular thread which is the most widely used, fatigue strength at 2×10^6 cycles is 59 MPa, which is considered here as the "fatigue limit" unless otherwise specified. A fatigue strength of 59 MPa is consistent with other published values [1, 2] for turned threads. Fatigue limit for positive buttress threads was nearly equal to that of the triangular threads, but the fatigue strength of trapezoidal and negative buttress threads were slightly higher. This may be attributed to the relaxation of stress concentration at the root. Machining a negative buttress thread, however, is rather difficult and, at the present time, lacks wide applicability because of the machining difficulty. In any case, the fatigue characteristics of a threaded connector cannot be significantly improved even if the type of thread is changed. If workability is taken into account, the triangular thread is an excellent compromise.

Figure 4 shows the effect of root radius on the fatigue limit. In this test, the root radius r was limited to 0.30-0.70 mm, and the conclusions drawn here may not be applicable to all cases. From this figure it is seen that the root radius has little effect on fatigue strength. Some variation is observed even for threads with the same raidii. In all cases, the fatigue limit is 59 MPa. The stress concentration at the root decreases with increasing root radius. However, if the root radius is increased, the rigidity of the threads is increased and localized contact with the internal threads is more likely to increase. It is

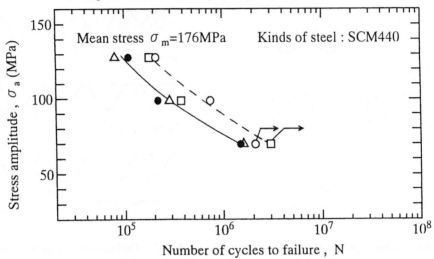

△ Triangular thread, φ 25, r=0.30 : dotted line
□ Trapezoidal thread, φ 25, r=0.20 ; dotted line
● Positive buttress thread, φ 25, r=0.30 ; solid line
○ Negative buttress thread, φ 25, r=0.40 ; dotted line

Fig. 3.   Effect of kinds of screw on fatigue strength

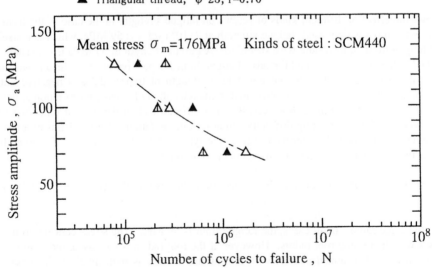

△ Triangular thread, φ 25, r=0.30 ; dash–dot line
△ Triangular thread, φ 25, r=0.50
▲ Triangular thread, φ 25, r=0.70

Fig. 4.   Effect of root radius on fatigue strength

considered that these effects cancel each other, causing a little change in the fatigue limit. Further tests using a wider variation in root radius are necessary to make further conclusions. If the root radius is changed, the tensile strength of the bolt may be decreased. Accordingly, an improvement in fatigue strength may only produce an adverse static effect and thread profile should be optimized by considering all factors involved.

The effect of bolt material on the fatigue strength is shown in Fig. 5. The tensile strength is increased by about 25% from 800 to 1,000 MPa by changing the material. In this case, there is a little difference in fatigue limit between the two materials. However, the fatigue strength of SNCM630 is lower by about one-fifth in terms of the number of cycles. It is widely known that the fatigue strength can be improved by increasing the tensile strength. However, the results shown in Fig. 5 are the opposite. The effect shown in Fig. 5 is attributed to two factors. The first is that the bolt is a kind of notched specimen. In ordinary fatigue, the fatigue limit of a plain specimens tends to increase with increasing tensile strength. However, even if the tensile strength is increased with decreasing notch radius, this difference in tensile strength does not have a noticeable effect on the fatigue limit [3]. Notch sensitivity also increases with increasing tensile strength and the fatigue strength of this specimen decreases more than for a plain specimen. The second factor is that of force transmission in the bolt though contact between the external threads and the internal threads. If the tensile strength is increased, the contact between the nut and bolt is apt to become one-side microscopically, although this contact seems to remain changed macroscopically. In other words, the effect shown in Fig. 5 is partly attributed to localized contact. If the tensile strength of the material is high, localized contact is not relaxed as the material does not yield.

Figure 6 shows the effect of changing nut material on the fatigue strength. The fatigue limit for S20C (69MPa) is a 17% improvement as compared to that for SNCM630 (59MPa). Furthermore, the fatigue strength is increased by about ten times in terms of the number of cycles. A nut will have a larger root diameter than the bolt and is subject to compressive stresses as compared to the tensile stress in a bolt. Accordingly, a considerable effect can be expected by making the nut material softer than the bolt material. Some researchers are of the opinion that the fatigue strength can be greatly improved by using a cast iron nut which has a lower elastic coefficient than carbon steel [3]. The use of cast iron nuts aims to equalize the load sharing between threads thus decreasing the internal peak stress in a bolt.

A New Method to Improve The Fatigue Strength of Bolts

The new method for markedly improving the fatigue strength of bolts for both pre-stressed and non pre-stressed applications has been developed. A typical shape of the optimized CD bolt is shown in Fig. 7. Shown here is the CD bolt with nominal diameter body, but the concept of the CD bolt is also applicable to bolts with a pitch diameter body and bolts with reduced shanks.

CD bolt is the trade name of a bolt manufacturer [4]. CD stands for "Critical Design for Fracture", and represents the best fatigue resistant profile of a thread known to the authors [5-7]. This design is not only effective for improved fatigue properties but also for delayed fracture.

Low fatigue strength in conventional bolts is attributable to four factors:
1) uneven load sharing among the threads,
2) concentration of tensile stress,
3) concentration of bending stress, and
4) localized loading.

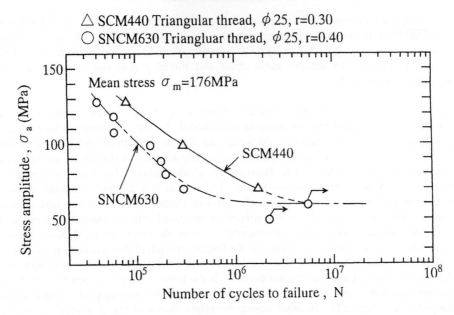

Fig. 5.   Effect of bolt materials on fatigue strength

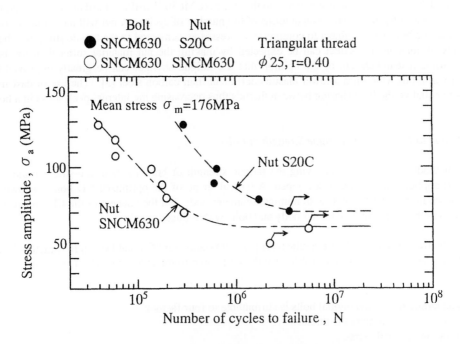

Fig. 6.   Effect of nut materials on fatigue strength

The CD bolt has been developed from the dynamic standpoint to improve the fatigue characteristic of bolts and incorporates all the measures necessary for the improvement of the fatigue strength of bolts. Accordingly, there is a marked difference in effect between the CD bolt and conventional bolts. The unique profile:

1) uniformly shares load between threads,
2) the height of the threads in engagement with the nut is decreased and the corresponding tensile stress concentration is reduced,
3) as the height of the thread is reduced, the distance between the loaded portion of the thread and the thread root is decreased and, if the threads are assumed to be cantilevers, the bending stress concentration at the root is reduced.
4) as the contact surface is increased, the contact area between the bolt and the nut is more easily deformed and contact is made between the bolt thread and the tip of the thread on the nut side which is more likely to be deflected. Accordingly, localized loading due to thread shape inaccuracies or other factors is reduced.

In other words, the CD bolt addresses all feaures which tend to lower the fatigue strength of traditional bolts.

Fatigue Strength of The CD Bolt

The fatigue strength of the CD bolt (2×10^6 cycles) is shown in Fig. 8 and is compared with the other bolts in Table 3. As shown in this table, the fatigue strength of the CD bolt is nearly double that of a bolt with triangular thread, 108 MPa vs. 59 MPa. In the pre-stressed condition, the CD bolt has an fatigue strength of 118 MPa vs. 88 MPa for a triangular thread. Figure 9 shows an example of analytical results by FEM about the difference of localized loading between the conventional bolt and the CD bolt. Though the result is not necessarily exact solution, the localized loading of the CD bolt becomes rather qualitatively smaller than that of the conventional one. As the CD bolt, particularly the threaded part, has a higher fatigue strength, fracture of the CD bolt occurs under the head where the fatigue strength is considered to be high, while fracture of the conventional bolt initiates at the end section of the nut (see Fig. 10). Figure 11 summarizes the factors affecting the fatigue strength of bolts and some possible countermeasures.

CONCLUSIONS

The main results obtained in this development are as follows:

1) Fatigue strength of screws is only marginally improved by changing the type of thread, root radius, and nut and bolt materials. The traditional triangular thread has an excellent total balance when considering static strength, machinability, and fatigue strength.
2) A new method for improving the fatigue strength of bolts has been presented. Fatigue limit for the optimized bolt is about twice that of conventional bolts for the stress amplitude. Improvement is observed whether or not a pre-stress is applied.
3) For a typical coarse threaded bolt of nominal size, the gradient of the thread part shaped by CD is about 6/100 and the incomplete thread is almost completely removed. In addition, the thread is connected to the body by a gentle arc of 10 mm radius. The optimum end section of the nut is such that about 70% of the CD-shaped part of the bolt goes into the nut.

Shin-ichi Nishida et al.

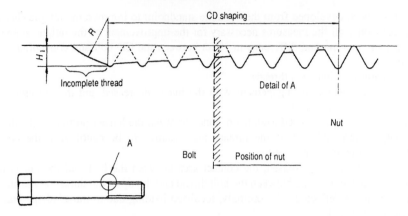

Fig. 7.   Typical shape of CD bolt [ref. Fig. 10(a)]

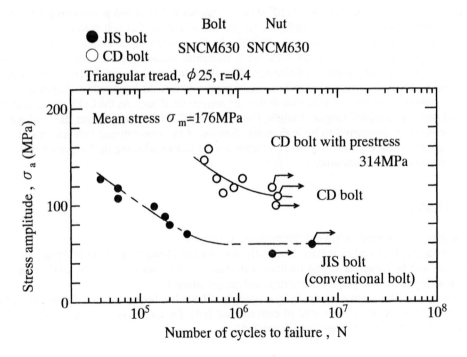

Fig. 8.   S-N curve for CD bolt

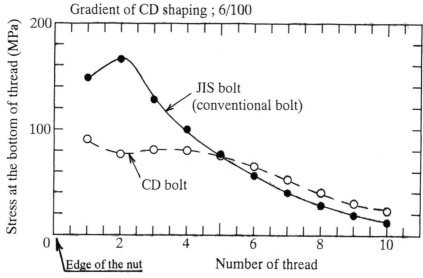

Analysis conditions ; MARC#28
8 nodes axial synmetry
236 elements
Gradient of CD shaping ; 6/100

JIS bolt
(conventional bolt)

CD bolt

Edge of the nut

Number of thread

Fig. 9. Difference of localized loading between conventional bolt and CD bolt (by FEM calculation)

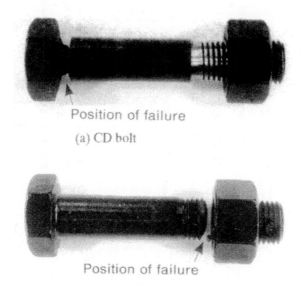

Position of failure

(a) CD bolt

Position of failure

(b) JIS bolt (conventional bolt)

Fig. 10 Difference in position of failure between

Table 3.   Summarized fatigue test results of bolts

Items for study	Type of bolt	Materials used		Size of bolt (mm)			Pretreatment	Test conditions		Fatigue strength at 2×10^6 cycles(MPa)	Increasing ratio of fatigue strength(%)	Remarks
		Bolt	Nut	Minor diameter	Root radius,r	Pitch		Mean stress (MPa)	Frequency (cycles/min)			
Fatigue strength of bolt with various threads	Triangular thread	SCM440	SCM440	25.0	0.30	3.0	No pretreatment	176	500	±59	100	Normal shape (standard bolt)
	Trapezoidal thread	SCM440	SCM440	25.0	0.25	3.0	ibid	176	500	±69	117	
	Positive buttress thread	SCM440	SCM440	25.0	0.35	4.0	ibid	176	500	±59	100	No effect
	Negative buttress thread	SCM440	SCM440	25.0	0.30	4.0	ibid	176	500	±69/±78	117/133	
Size effect	Triangular thread	SCM440	SCM440	32.0	0.80	3.6	ibid	176	350	±59	100	
	ibid	SCM440	SCM440	40.0	1.00	5.5	ibid	176	350	±59	100	
Effect of root radius r	ibid	SCM440	SCM440	25.0	0.50	3.25	ibid	176	500	±59	100	No effect
	ibid	SCM440	SCM440	25.0	0.70	3.5	ibid	176	500	±59	100	No effect
Effect of mechanical properties of bolt	ibid	SNCM630	SNCM630	25.0	0.40	3.25	ibid	176	500	±59	100	No effect Slightly negative effect
Material of nut	ibid	SNCM630	S20C	25.0	0.40	3.25	ibid	176	500	±69	117	
Effect of prestressing (new method)	ibid	SNCM630	SNCM630	25.0	0.40	3.25	Prestress 421MPa	176	500	±88	150	
	ibid	SNCM630	S20C	25.0	0.40	3.25	Prestress 363MPa	176	500	±73.5	125	
	Gradual cut-off of bolt's thread ibid	SNCM630	SNCM630	25.0	0.40	3.25	Prestress 323MPa	176	500	±118	200	CD bolt
Effect of gradual cut-off method of bolt's thread (new method)	Gradual cut-off of bolt's thread*	SNCM630	SNCM630	25.0	0.40	3.25	Gradual cut-off from 1st thread to 8th one	176	500	±108	183	Engaged nut at 4th thread

* Gradual cut-off of bolt's thread means CD bolt.   † Fatigue life becomes about 1/10th

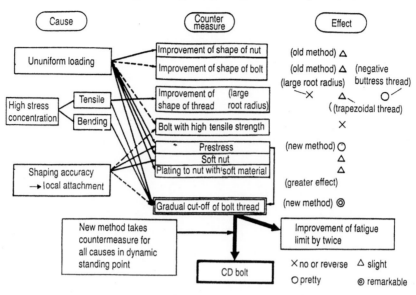

Fig. 11. General factors governing the fatigue strength of bolts, countermeasures, and their effects.

REFERENCES

1. Machine Design Handbook Editing Committee (1973). Machine Design Handbook, Maruzen, pp.969.
2. Ishibashi, T.(1954). Prevention of Fatigue and Failure of Metals, Yokendo, pp.256-275.
3. Otaki, H. (1981). Machine Design, **25**, pp.27-32.
4. Nippon Steel Bolten Co. Ltd. (1983). A brochure of Nippon Steel Bolten, CD bolt, pp.1-4.
5. Nishida, S. (1992). Failure Analysis in Engineering Applications, Butterworth Heinemann Co. Ltd., pp.68-104.
6. Nishida, S. (1982). Maintenance, **30**, pp.33-40.
7. Nishida, S. (1982). Maintenance, **31**, pp.40-48.

Author Index

Author Index

Printed and bound by CPI Group (UK) Ltd, Croydon, CR0 4YY

08/05/2025

01864846-0001